THE CONCEPT AND THE ROLE OF THE MODEL IN MATHEMATICS AND NATURAL AND SOCIAL SCIENCES

SYNTHESE LIBRARY

A SERIES OF MONOGRAPHS ON THE

RECENT DEVELOPMENT OF SYMBOLIC LOGIC,

SIGNIFICS, SOCIOLOGY OF LANGUAGE,

SOCIOLOGY OF SCIENCE AND OF KNOWLEDGE,

STATISTICS OF LANGUAGE

AND RELATED FIELDS

Editors:

B. H. KAZEMIER / D. VUYSJE

THE CONCEPT AND THE ROLE OF THE MODEL IN MATHEMATICS AND NATURAL AND SOCIAL SCIENCES

Proceedings of the Colloquium

sponsored by the Division of Philosophy of Sciences of the

International Union of History and Philosophy of Sciences

organized at Utrecht, January 1960,

by Hans Freudenthal

D. REIDEL PUBLISHING COMPANY / DORDRECHT-HOLLAND

GORDON AND BREACH / NEW YORK

SOLE DISTRIBUTORS FOR NORTH AND SOUTH AMERICA
GORDON AND BREACH
SCIENCE PUBLISHERS, INC.
150 Fifth Avenue, New York 11, N.Y.

Printed in the Netherlands by D. Reidel, Dordrecht

CONTENTS

LEO APOSTEL

TOWARDS THE FORMAL STUDY OF MODELS IN THE NON-FORMAL SCIENCES

SUMMARY

I. The function of models in the empirical sciences.

II. Structure and purpose: conditions of a structural nature which models should satisfy in order to accomplish their function.

III. Generalisation and specialisation of the classical definition of model, in view of the above requirements:
(a) the algebraic model concept
(b) the semantic model concept
(c) the syntactical model concept.

IV. Attempt towards reunification: the concept of model on a pragmatic basis.

I. THE FUNCTION OF MODELS IN THE EMPIRICAL SCIENCES

Scientific research utilises models in many places, as instruments in the service of many different needs. The first requirement a study of model-building in science should satisfy is not to neglect this undeniable diversity (as has sometimes been done) [1]), and, when recognising this multiplicity, to realise that the same instrument cannot perform all those functions (often the multiplicity of function is recognised but either not to a full extent, or not with respect to the difference of structure it implies) [2]). We are going to mention some of the main motives underlying the use of models:

(A) For a certain domain of facts, let no theory be known. If we replace our study of this domain by the study of another set of facts for which a theory is well-known, and that has certain important characteristics in common with the field under investigation, then we use a model to develop our knowledge from a zero (or near zero) starting-point. This is what happens in neurology: we replace the central nervous system by a digital or anologue computer showing certain of the neurological peculiarities, and study this new object.

(B) For a domain D of facts, we do have a full-fledged theory, but one too difficult mathematically to yield solutions, given our present techniques. We then interpret the fundamental notions of the theory in a model, in such a way that simplifying assumptions can express this assignment: under these simplifying assumptions, the equations become soluble. Using the theory of harmonic oscillators in the study of heat conduction is an example of such a procedure.

(C) If two theories are without contact with each other we can try to use the one as a model for the other or to introduce a common model interpreting both and thus relating both languages to each other.

(D) If a theory is well confirmed but incomplete, we can assign a model in the hope of achieving completeness through the study of this model. Special cases of this procedure are: a qualitative theory is known for a field and the model introduces quantitative precision; or a quantitative theory is used for a field, but not securely established, and the model circumscribes the solid core of the theory in qualitative terms.

(E) Conversely, if new information is obtained about a domain, to assure ourselves that the new and more general theory still concerns our earlier domain, we construct the earlier domain as a model of the later theory and show that all models of this theory are related to the initial domain, constructed as model, in a specific way.

(F) Even if we have a theory about a set of facts, this does not mean that we have explained those facts. Models can yield such explanations (particle or wave theories of light, or statistical mechanics, are important examples of explanation through model building).

(G) Let a theory be needed about an object that is too big or too small or too far away or too dangerous to be observed or experimented upon. Systems are then constructed that can be used as practical models, experiments on which can be taken as sufficiently representative of the first system to yield the desired information.

(H) Often we need to have a theory present to our mind as a whole for practical or theoretical purposes. A model realises this globalisation through either visualisation or realisation of a closed formal structure.

(I) It often occurs that the theoretical level is far away from the observational level; concepts cannot be immediately interpreted in terms of observations. Models are then introduced to constitute the bridge between the theoretical and observational levels, the theoretical predicates being interpretable as predicates of the model and the observational predicates being also interpretable as predicates of the model, the model furnishing lawful relationships between the two interpretations. This intermediary model can be used to construct the abstract theory or, once it exists, to find for it domains of application.

We are certain that still other functions could be found for models in empirical research. We are also certain that with the help of some supplementary assumptions, some of these functions of models could be reduced to some others. Still, it is true to say that the aims mentioned – theory formation, simplification, reduction, extension, adequation, explanation, concretisation, globalisation, action or experimentation – constitute a kind of system. It appears indeed that models have been introduced in function of relations between theories and theories, between experiments and theories, between experiments and experiments, between intellectual structures and the subjects using these structures, and in all these cases this has occurred in order dynamically to produce new results, or in order to tie up new ones with old ones as guarantees, or simply to establish relation.

Most of these cases have occurred in the list above; while it is certain that the problem-solving behaviour of man knows other factors than those mentioned here, it seems to be true that the model as a tool mediating between some of these factors has here been adequately localised.

What are now the questions we wish to ask about the model-concept in these various roles? Among others, the following:

(i) Is it possible to derive from the description of the function to be fulfilled, the features a model should have to achieve this purpose?

(ii) Will the type of model needed to fulfil a given function have differing structure for theories, facts or actions of different types?

(iii) Can conditions be formulated determining when models can fulfil one of these functions, and when they cannot do so, when they are the only instruments or possible instruments among others?

(iv) Can some common feature be distinguished, either among the various aims, or among the various eventual structures, thus unifying to some extent the family of models?

The importance of these problems is clear. The concept of model will be useless if we cannot deduce from its function a determinate structure. The scientist in his comments uses the 'model' concept in all the ways we described and thus discards the simple and clear language of model-theory in formal semantics or syntax. Can we give to his use of the term an adequate rational reconstruction, or are we prevented from doing so? We are convinced that it is possible to derive structural features from the functional characterisation, and it is to this attempt that we now immediately proceed.

II. STRUCTURE AND PURPOSE

Let then R(S,P,M,T) indicate the main variables of the modelling relationship. The subject S takes, in view of the purpose P, the entity M as a model for the prototype T. We saw above a classification of possible purpose, S of values for P. Let us now only mention classifications for M and T.

Model and prototype can belong to the same class of entities or to different classes of entities. The following possibilities immediately offer themselves:

M and T are both images, or both perceptions, or both drawings, or both formalisms (calculi), or both languages, or both physical systems.

All these possibilities have occurred. But we can also have the heterogeneous case: M can be an image, T a physical system, or inversely; M can be an image and T a perception; M can be a drawing and T a perception; M can be a calculus and T a theory or language; or inversely. M can be a language and T a physical or biological system.

Among each of these classes, a finer subdivision could and should be considered. If the model is a theory, this theory can have all degrees of systematic unity, or of completeness, or of confirmation or confirmability; if the model is an image it can have all types of organisation, of vagueness, of closedness.

Will there be an interaction between the multiplicity of values for P and the multiplicity of values for M and T? There can be no doubt about this. Can the model-prototype relationship that exists in formal semantics teach us anything about similar relations occurring in domains so widely different? It is, once more, our conviction that it can do so. More formally: if L is the relation between M and T, then we claim that from R (S, P,

M, T) we can derive facts about L(M, T), depending upon the different values of P, M and T.
To substantiate this claim, we will now analyse some special cases.

A. *Models and the progress of research*

Each single purpose, among those we have mentioned, is in itself ambiguous. It should not astonish us, then, that a model could aid the progress of a scientific system in many ways: the science in question could need only completion, or, alternatively, restructuration. If the science in question is to be completed, then the model used to lead it to its completion should have properties not mentioned in the initially existent science. The study of these properties could then yield completion. For this to occur, a multiplicity of non-isomorphic models should exist, and this is a perfectly normal case, in the formal sciences. This type of progress through model construction has the following two limitations: (a) it cannot furnish transformation, but only addition of new details; (b) it is intrinsically limited (when the final description of the model is completed, the process must stop). Those who wish to use model construction to rebuild their discipline, or those who wish to guarantee indefinite evolution, should use another model concept. But let us for a moment restrict ourselves to the more modest task one could hope to achieve, when looking only for completion and not for restructuration. Why should we construct models to reach this aim? Why can we not simply consider the possible hypotheses we could add to our theory, consistent with the already accepted ones? Why should we use this devious procedure while a more direct one is at our disposal? If we compare, for a given language L, the set of possible complete languages L1 .... Ln obtained through addition of supplementary hypotheses to the set of models of L, then this set of models should have in some sense a structure that makes selection between the models easier than selection between the formalisms.

In principle, using the classical concept of model, the set of complete extensions of L and the set of models of L should be isomorphic. But if this is the case, then, in principle, models could be dispensed with. What then should be the concept of model that could make the use of models indispensable for the aim of simple achievement of a theory already started? Either there should be fewer models than possible additions of hypotheses, or the relations between the models should be simpler than the relations between the consistent extensions. In the first case there should be stricter demands on the model concepts than in the formal sciences, in the second case, the distinctions between models should be clearer, and so again we should have requirements preventing some intermediary cases allowed by the classical model definitions, from appearing among the list of models.

This being known, let us now ask how model-building could advance restructuration of a theory. If the so-called model is not really a complete model in the classical sense but only satisfies certain best-confirmed or most-used laws of the theory, then the model, not satisfying certain other less central features of the theory, could help us in replacing these by others that would be satisfied by the model. This type of partial correspondence and partial discrepancy between model and theory could eventually lead to indefinitely continuing development.

But this is not the only way in which models could help towards restructuration.

Let us suppose that we use a series of partial models, each of them representing part of the theory to be modelled, but none of them satisfying it as a whole, and some of them inconsistent with each other. Research into the extensions of these partial models that would include a maximum number of other partial models could equally lead to reformulation of the initial theory.

Or let us use a multiplicity of complete models simultaneously; or a combination of complete models and partial models. In such a scheme arbitrariness of the exact selection of the entity representing a concept will lead to search for new requirements that will yield a non-arbitrary selection.

A limit case of this situation is the use of a locally inconsistent model. In classical Rutherford atomic theory, it was clearly recognised that the nucleus of an atom should explode under the electromagnetic laws of the time (due to the internal repulsion of the positive charge). So everybody knew the model to be inconsistent. But this inconsistency was accepted because the nucleus could, in the applications where its charge was needed, be treated as a point, and where its dimensions were needed, its charge and the internal properties of it did not intervene.

A final possibility to help restructuration is the use of undefined or vague models, the indefiniteness of which suggests and allows completion in given directions.

We do not believe that this exhausts all possibilities, but it seems to us that the possibilities mentioned here form a systematic whole: models are used for system restructuration because of their relations with the system (partial discrepancy); because of their relationship among each other (partial inconsistency, at least multiplicity); because of their relationships with themselves (locally inconsistent or locally vague). To summarise: models used for completion should satisfy more stringent requirements than the classical model in the formal sciences, while models used for system restructuration should simultaneously satisfy more stringent requirements (our inquiry as to the conditions that make the detour through model-building desirable remains valid in this last case), and more lenient ones (partly inadequate, vague, multiple and locally inconsistent models).

Intuitively it thus appears that, at least for one of the possible aims of model-building, the bridge between the formal and the functional exists. Can we now build a formal theory about approximate, partial, multiple, locally inconsistent, or vague models? To inquire about this problem will be the task of our third section.

B. *Models and the initiation of research*

Let us have a set of data about a domain, either very irregular and complex, or very incomplete. We wish to build a theory. We could try to tackle the data immediately themselves. But if we have some reason to suppose that they are grievously incomplete, or that they are very complex functions of the really independent variables, the following strategy seems fruitful: select one very specific law of the domain, try to build a mechanism, a model that satisfies this very specific law, and then, in view of this model, localise the form or structure of our data – the way in which they are complex and what supplementary data should be sought after if our initial ones are given functions of our model concepts.

In order for this method to be fruitful, the basic law we represent in the model should be such as to have very few models satisfying it in a given range, or, if many satisfy it, should be such that in all those cases, the model of the data derivable from the model of the law allows analysis of these same data into entities of simpler and more regular structure.

Compared to the classical model situation, we have here a structure that should satisfy a requirement of correspondence with a theory T, of very elementary character, but should moreover satisfy a requirement of correspondence with one partial model of the elementary theory: namely the data, and this second requirement is such that the mapping of D into M should give an image M (D) having a more regular structure than D. The triangular relation between T, M and D, L (D, T) being itself approximately true is thus here an essential feature of the situation. We saw in (A) that for the needs of research composite models should be considered; now we see, in a symmetrical fashion, that other needs suggest the study of models of complexes of theories, ordered in certain ways.

C. *Models and experience*

Already in his 'Introduction to Semantics' [3]), Rudolf Carnap makes the distinction between a logical and a descriptive interpretation of a calculus. Without claiming that an empirical science is or can be a calculus in Carnap's sense, we should consider the considerations introduced in making this distinction. If we give for all signs of a calculus rules of designation, or, if we give for all sentences of a calculus rules of truth, we give an interpretation of this calculus. On pp. 203-204 Carnap stresses that for application it is necessary to construct a bridge between 'the postulate set and the realm of objects' (p. 204) and that this is called 'constructing models or giving interpretations' (phrases he uses synonymously). An interpretation is a true interpretation if whenever a sentence implies another in the calculus, in the interpretation whenever the first sentence is true, the second is equally true, and whenever a sentence is refutable in the calculus, it is false in the model. Such a true interpretation is a logically true interpretation, if the sentences that become true, become logically true. An interpretation is a factual interpretation if it is not a logical interpretation. An interpretation is a descriptive interpretation if at least one of the undefined signs of the calculus becomes in the interpretation a descriptive sign, and while Carnap gives, pp. 58-60, clear examples of descriptive signs (names of single things, of observable properties), he stresses on pp. 59-60 that no general solution in general semantics is known for the problem of distinguishing between logical and descriptive signs. These definitions are important for us, because it is clear that the concept of model in the empirical sciences, when it is used in the following context – 'the world is a model of our sciences, in so far as these sciences are true' (or conversely the aim of science is to construct a calculus for which reality is the only model) – takes the concept 'model' in the sense of a factual and descriptive true interpretation. Here we realise that we are up against some of the main problems of recent logical research: the definition of logical truth and the definition of descriptive sign. But there are here still more problems that could not easily be treated by Carnap in 1946 but that have become especially prominent: it has been recognised that most calculi have many more models than they were intended to have (the existence of non-standard models is a case in point). When now we talk about 'models' in empirical sciences, we mean, if we want reality to be a model of our science, to talk about an intended model. The only writer who has tried until now to introduce a general distinction between logical and descriptive constants, and to formalise some of the properties that distinguish intended from non-intended models, is Kemeny [4]). In the sense we are discussing here, a model in the empirical sciences is an intended factually-true descriptive interpretation. (Or, in some other contexts: a non-intended arbitrary interpretation, used to clarify such intended factually-true descriptive interpretation.) If we now introduce, with Kemeny, as definition for logical truth, validity in all interpretations, and the property of being a descriptive constant as not being assigned the same value in all interpretations, and if moreover we accept (again from Kemeny) that all models are interpretations that have the same domain of individuals as the intended one but other assignments for non-logical constants, then, if we are to study models for empirical sciences, we must study sets of structures having the same individuals, and varying for all undefined constants, their assignments in the models (except for the classical logical constants), and not differing from at least one among them otherwise than through this variation.

It is clear that we could very well consider other definitions for logical truth, or for descriptive constants (e.g. not completely definable or applicable without ostensive definition), but our claim here is that, once general definitions for these key terms are provided, a formal structure is given to the model concept in its function as relater of theory and experience, formal structure that could be studied.

Let us however stress one more feature about the semantics of the empirical sciences: in the formal sciences model-building signifies mapping a calculus upon a fragment of set-theory. A formula F is true if there exists a set showing between its members certain relations. In order to apply this method to sentences taken from the empirical sciences, we should there also select some basic science in the language of which the truth conditions for the sentences of the empirical sciences could be formulated. But (a) no empirical science could fulfil this function at present, and (b) it is doubtful if set-theory could again fulfil its old function in this context. Certainly the tendency exists to reintroduce set-theory for the semantics of the empirical sciences and to have it serve these new needs. We can say that a system satisfies a given law L on certain variables, if the set of numbers representing measures of these variables exhibits a relationship derivable from a set of initial conditions I and from the law L.

What happens in this definition is that we define M as model of T, if there exists a structure N, standing in a certain relation to M, and if N is homomorphic with the elements of a class K of models of T, with respect to given predicates. The structure N is the set of measuring results, on M, the class K of models of T is the class of models in the classical sense of the theory, sharing certain initial conditions, and the predicates are the ones corresponding to the variables measured. This is already in considerable deviation from the classical use of the model-concept, though definable with respect to it; but the essential departure is that here the model, at the limit, becomes a structure for which the propositions of the formalism are verified (not: are true). Semantics, which in the realm of formal languages was the foundation of theory of confirmation itself, here rests on the theory of confirmation (the more so, if we realise that we should add that the initial conditions under which the system obeys the law should be either the true initial conditions, or the verified or highly-confirmed ones).

We should thus confess that either we must accept this consequence, and thus define first 'theory of measurement' and 'confirmation', and only later define 'model' and 'truth' for empirical sciences, or, instead, select for the empirical sciences a basic language that could be used here in the same way in which set-theory is used for the formal sciences. The search for such a basic language should not be arbitrary, because set theory has a very specific place among the formal sciences; this place should be defined, and once this is done we could look for the discipline that has an analogous place among the empirical disciplines.

Let us now make a final point: as in the immediately preceding paragraph, we see the need to study the relationship between the model and a complex structure of theories. If the model should be the intermediary between either theory and experience, or theory and reality (we saw that in both directions we have a different modelling relationship), then this intermediary character creates another problem. Let us say that a model is an intermediary between two formalisms if it is a model of both, and if the properties in the model that relate it to T1 are, in M, related to the other properties that relate it to T2. The number and kind of these relations could even be used to develop the notion of 'degree of intermediacy'. Even if we want to avoid the problems of descriptive interpretation, we can study the properties of models of such a combination of theories.

To summarise: the problem of descriptive interpretation, the problem of the empirical basic language, of the definition of interpretation through measurement and confirmation, and the final problem of intermediacy, show us that here, also, structure and purpose are related – definable with reference to the formal concepts but not identical to them.

D. *Models and experimentation*

We cannot afford to lose an airplane each time we wish to see if it is able to resist under certain velocities. Therefore we build model airplanes, that we test on model velocities or pressures. If the model is adequate we should be able to derive from these model experiments the desired information. Essentially we have changed scale and have tried to leave everything else invariant. The difficulty is that when I change scale I always change something else; the problem is how to correct for the changes introduced through the scale change, or how to find a series of variables that are perhaps in their relations affected by the scale change, but not with reference to the relation we are interested in.

Two similar triangles have their sides in the same proportion even though their absolute magnitudes may be extremely different. We can generalise this concept of analogy, or proportionality, so important in Greek mathematics, and say that if a physical system is completely determined by *n* dimensions, as the triangles are by their sides, one system is a model for another system if the relations between these dimensions remain the same, even though their values are changed. As mentioned before, I cannot hope to reach absolute similitude among physical systems, but I can hope to reach corrigible dissimilitude, or approximate similitude. To generalise now (and not to let ourselves be tied down by the scale factor) let us say that in order to construct models for action, there must be at least one variable among those that determine the system that may be either arbitrarily varied, or at least varied in a wide range, without modifying, or without modifying too much, the relations between the other determining variables of the system. Let us try to show in general that structural properties can be derived from this demand. Let x be a function of two variables y and z. Let the three variables be quantitative variables. When will the value of x be independent from variations in the values of y and z? If x is an increasing function of y and a decreasing function of z, and if the increase produced by an increase of y is exactly equal to the decrease produced by an accompanying increase of z (z and y being inversely related) then x will remain invariant. If we want such a function in Boolean algebra (i.e. in an elementary fragment of set-theory, the general foundation of model-theory) we can look at the function Un. (Int(x, Cy), Int (y, Cx), where Intersection indicates the common part of two classes, Union the sum of the two classes, C the complement of a class and where x and y are used here as variables for classes).

This definition for a function of two variables can naturally be extended for n variables, and the quantitative nature is by no means needed, as shown in the example from Boolean algebra. If our variables were relations, we could construct the same example. We could now give in general the following definition: in our present sense of model, M is a model of T if both are relational structures and if the relations of both are invariant functions of the relations they do not share. The only new concept used here is the concept of 'invariant function' (and this concept, as stressed, is easily definable in general semantics).

But now that we have reached this very structural and very special-looking concept (certainly much more demanding than the classical case), let us remind ourselves that we do not need to experiment upon a system if we know it completely. Not knowing a system completely, the form of all laws it obeys are *a fortiori* not known to us either. The model concept we are then compelled to use is an approximation in two stages:

(a) if a set of equations is given, and if these equations are invariant under certain transformations, then two systems are models for each other if one of them satisfies the system of equations and if the second system can be produced by applying to the first some transformations under which the system equations are invariant.

Nothing implies that the system of equations is thus a complete equation system of the system.

(b) an approximation going in the other direction is this: let us know *a priori* what are the variables the knowledge of which completely determines the system, even though we do not know how this determination occurs, and what is the exact form of the equations.

Let there be given the way in which these variables depend upon the fundamental variables of our science. Then using these expressions, we shall say that M is a model of T if both have the same dimensions (here the defined variables should depend for all their values upon the undefined ones in a similar fashion).

The first approximation to physical similitude forces us to ask: what type of transformations should we consider and how should we measure the approximation to completeness. And the second approach forces us to ask the question: how can I, without knowing the function that relates x and y, say that x depends upon y and how can I, from assertions about dependence or independence alone, infer the form of this dependence? Dimensional analysis has examined these and similar problems for many special cases [5]), but the essence of our task here is to show that the same questions should be asked for relations in general, as a part of the study of the specific model concept that is used in experimental action. Our earlier remarks about the general invariance of functions shows that independence can be defined in a structural fashion. The formal problem is the following: if I have a sequence of variables and relations among these variables, what circumstances make modelling-experiments possible or impossible? If I have x, y and z as variables and if the following relations all hold $F1(xy)$, $F2(xz)$, $F3(yx)$, $F4(yz)$, $F5(zx)$, $F6(zy)$, and if moreover $F7(xyz)$, and if moreover higher order interactions $F8(x, F1(yz))$ exist, then what type of modelling-experiments become possible if successively either some of the higher order interactions between relations and variables are eliminated (the dependence $f(xy)=g(y)$ is the prototype of an obstacle against modelling), or some specific conditions are introduced? In view of this question, we reach the following definition of model: if two systems, according to dependence-independence tests, are determined by the same fundamental variables, which, according to a given dimensional analysis, are determined by a set of equations wherein a certain kind of interactions does not occur, then those two systems are models for each other.

Let it then here be said that the double relativity of the dependence tests that are not complete descriptions, the dimensional analyses that are not unique, and the syntactical characteristic about the existence of laws of certain forms, makes this model-concept extremely specific, both more severe and more lenient, as always, than the one we know so well from classical semantics.

E. *Models and explanation*

Models are given as explanations of the systems they are models of. Why should models be needed for this purpose and how can they explain? There is perhaps no clearer refutation of the so often heard thesis according to which to explain is to infer, than the fact that explanation occurs so often, or even nearly always, through model-building.

The definition of explanation is once again one of the unsolved problems of the philosophy of science. It is thus very difficult to determine the structural properties a model should possess in order to be able to explain. We only want here to propose our personal hypothesis, without claiming more than plausibility in its favour. If we look at the history of science, we see that, in physics at least, two major explanatory models have been dominant: the atom model and the field model, the discontinuous and the continuous, the pluralistic and the monistic, notwithstanding the fact that neither atoms nor fields are familiar or simple entities. In the theory of gases, in the theory of light, in cosmogony, in the theory of electromagnetism, in nuclear physics at the present moment, these two explanatory models have always been influential. To state this fact more formally: physics seems to try to reduce all law either to the laws of Newtonian mechanics, or to the laws of Maxwell's electromagnetism, and, if possible, to both. Can we extrapolate towards the future, or towards other sciences? We could try in various ways to understand this tendency: if explanation is the derivation of the observed facts (always presenting a mixture of foreground atomism and background continuity) from premisses at a maximum distance from these observed facts, then we could claim these to be the two extremes. If explanation is analysis, then we could try to prove that the two extreme poles of analysis are the reduction of all plurality to the unity of the field whose features will have to be responsible for the observed universe, or the reduction of all order and unity to the pluralism of the particle on whose disorder our order is to be built. If to explain is to make anthropomorphically understandable, the world of the discontinuous is the world of the tool, and the world of the field is the world of the environment. We can only offer these suggestions as guesses; our only excuse is that nobody seems to have better ones. At least these guesses explain why we need model-building to explain. In terms of this hypothesis, an explanatory model is representation of a theory in the theory of complete or partial differential equations. The philosopher of science should certainly describe in more general terms what distinguishes these two theories from other ones, in order to understand their privileged position. But even before undertaking this task he can state that approximate models will have to be introduced in order to provide for systems having a very different structure a model in terms of these differential concepts.

Once more, we find a very specific addition to the classical theory of semantics, and also a very specific generalisation.

F. *Simplification and model-building*

It is rather paradoxical to realise that when a picture, a drawing, a diagram is called a model for a physical system, it is for the same reason that a formal set of postulates is called a model for a physical system. This reason can be indicated in one word: simplification. The mind needs in one act to have an overview of the essential characteristics of a domain; therefore the domain is represented either by a set of equations, or by a picture or by a diagram. The mind needs to see the system in opposition and distinction to all others; therefore the separation of the system from others is made more complete than it is in reality. The system is viewed from a certain scale; details that are too microscopical or too global are of no interest to us. Therefore they are left out. The system is known or controlled within certain limits of approximation. Therefore effects that do not reach this level of approximation are neglected. The system is studied with a certain purpose in mind; everything that does not affect this purpose is eliminated. The various features of the system need to be known as aspects of one identical whole: therefore their unity is exaggerated. When isolation, clarity, unity, closedness, essentiality and homogeneity of perspective and viewpoint are reached, an adequate model either expressed in equations or in a drawing is formed.

Moreover, both in the verbal and in the pictorial model, we represent either parts of the system and their connections, or states of the system and their connections, or simultaneously parts and states and their connections. The aim of simplification and globalisation will however be more accurately served through concentration on one of our two main sub-goals. Let it be clear that here the model should not be richer than the system it is a model of, but poorer. The model in the service of the progress of science should be more complex; this one should be more completely focused. It is in this sense that the ethical meaning of the word model meets the epistemological meaning. For certain ethical systems (not for all) the ideal man is the model of man, in this sense of the word model (an exaggeration of idiosyncrasy, a clarification of internal structure). We think that it is easy to recognise that the features of elimination of certain predicates, elimination of certain parts, closure for parts and states, regularisation of the overall structure, are structural demands that can be defined for very general relational systems in a truly general semantics. The importance of isolating this type of model concept lies in the fact that it cuts through the opposition of image and word, and explains why (as so often in economy) a system of equations is called a model, where a few pages earlier or later, a picture had the same attribute.

Here we want to close our review of the structural correspondents of our functional characterisation for models in the empirical sciences. We claim that we have made it plausible that a thorough analysis of the different aims of model-building shows us that very definite structures are needed to achieve these aims, and moreover (and this is centrally important) (i) *that these structures depart from the classical concept of model in many different ways but* (ii) *that they can be studied and ordered, using this same classical concept of model as a centre of perspective.*

We now want in our third section to show that if we start from the classical concept of model and if we apply to it certain natural operations of strengthening and weakening, we reach from the opposite direction the structures we tried to define from a functional point of view before. We shall thus be able to a certain extent to show that the science of formal semantics could still fruitfully be studied with reference to these more general models.

III. CLASSICAL AND GENERAL MODELS

Conceptually, if model and prototype are both general systems or structures, we have the most abstract case. If we impose on one of the terms of the relationship the demand that it should be a language, we have a more special case, and if both terms should be languages we have the most special case. The natural order to study generalisations of classical model concepts is to take the algebraical model concept first, to proceed afterwards to the study of the semantic concept of model (relating a language to an arbitrary domain) and to finish finally with the syntactical concept (relating two languages to each other).

A. *Algebraic models*

Our purpose in this section will be to define several approximate or strengthened forms of isomorphism. To reach this aim let us first realise what is included in the notion of isomorphism.

Two sets D1 and D2 are isomorphic with respect to relations R and S, defined respectively on D1 and D2 at least, if the following situation occurs: there exists a mapping function F such that to each member of D1 there corresponds one and only one member of D2 under F, and if moreover whenever members of D1 stand in the relation R, their F-images stand in the relation S, and inversely, then we say that the two domains are isomorphic under the two relations R and S. Two sets will be completely isomorphic if they are isomorphic under all their relations. Two relations will be completely isomorphic if they are isomorphic on all their domains. Two sets will be called isomorphic with reference to a class K of relations if the relations under which they are isomorphic have to belong to the class K. Two sets will be called isomorphic under a set K of relations if the mapping relations that correlate the relations on the two sets have to belong to a given class K.

A well-known and much-used generalisation of isomorphism is homomorphism. Here the correlator has not to be one-one but is allowed to be many-one. As for the rest, the relations remain the same.

We wish to consider approximative isomorphisms and homomorphisms and take some steps towards ordering or even measuring these approximations.

We give some possible forms of approximation.

Approximation I: The correlator may be such that this function does not map the whole of the domain of R on S, nor the whole of the domain of S on R. We then have an approximate direct or inverse correlator. A correlator A will be a closer approximation than a correlator B if the class of members of the field of R that is not mapped upon the field of S by A is a proper subset of the class of members of the field of R not mapped upon the field of S by B. If we possess a measure on our sets we can define one approximation to a correlator as better than the other one, if the measure of the non-mapped set is smaller in the first case.
Approximation II. The correspondence may be such that not always when R exists in D1 between some elements, S exists in D2 between the images of these elements. The correspondence C_3 is a closer approximation than the correspondence C_2 if for C, the *n*-uples where the images have not the corresponding relations are a proper subset of the set of *n*-uples which for C_2 do not have the corresponding relations.
We shall say that an App. II-neighbourhood-relation isomorphism exists if for all cases in which the correspondence does not hold a relation lying in the neighbourhood (provided such a concept is defined) of the relation that should occur, holds. We shall say that an App. II-neighbourhood-element isomorphism exists if for all cases in which the correspondence does not hold, some elements in the neighbourhood of the image elements present the desired relation.
We wish to stress that approximation-neighbourhood isomorphisms of type I (both of relation and element kind) can be defined in the same way. If the definition of neighbourhood for one approximation is a refinement of the definition of neighbourhood for another approximation, then the first is a closer neighbourhood element or relation approximation than the second.
Approximation III. An App. III-isomorphism is both an App. I and an App. II isomorphism.
Approximation IV. In all preceding approximations we have considered the set D and the relation R to be classical sets or relations, defined everywhere, and precise everywhere. Let us now introduce for sets and for relations indetermination domains; i.e. elements for which it is not decidable if they belong to a set or not and couples for which it is not decidable if they belong to a relation or not. D'1 is a closer approximation to D1 than D''1 if the determination domain of D'1 has a larger inter-

section with D1 than that of D″1 (if no measure is available: if the first intersection includes the second). D′1 is moreover a tighter approximation to D1 than D″1 if the indetermination domain of the first is included in the indetermination domain of the second. Finally D′1 is a more adequate approximation to D1 than D″1 if the intersection of the indetermination domain of D′1 with D1 is included as a proper part in the intersection of the indetermination domain of D″1 with D1. It is important to understand that a closer, a tighter and a more adequate approximation to the same set are by no means always identical. Moreover, we have supposed here the set D1 to be exact. It is obvious that we should also consider approximations to inexact sets.

A relation can have an indetermination region for its domain, and for its co-domain. If we want to consider isomorphism between approximate sets and approximate relations, we meet first the problem of the intersection of the indetermination region for relations and for sets. Either one of both can be zero, or both can exist; if they both exist they can intersect (even be included in each other) or be disjunct from each other. It is obvious that the case in which they coincide will be the easiest one for our purpose.

Let us say that if we have a set with an indetermination domain, and on this set a relation R with again an indetermination domain, then there is exact isomophism between both if there is a correlator mapping members of the kernel of D1 on members of the kernel of D2, members of the indetermination domain of D1 on members of the indetermination domain of D2, and such that if R holds between elements of D1, S holds between elements of D2 and if a *n*-uple is in D in the indetermination domain of R, the images of it are in D2 in the indetermination domain of S. This strict isomorphism immediately gives birth to a series of approximations: we may call alpha IV approximation the case where the correlator maps elements of the kernel of D1 on elements of the indetermination domain of D2 (or inversely), beta IV where elements of indetermination domains are mapped on elements of kernels, gamma IV where elements of indetermination domains are not mapped (and thus in a sense mapped on complements), delta IV approximation where *n*-uples in the kernel of R are mapped on *n*-uples in the indetermination domain of S.

We shall call one isomorphism closer or tighter or more adequate if a closer, tighter or more adequate approximate to both D1 and R is correlated with a closer, tighter or more adequate correlate to both D2 and S. This definition again gives us the occasion to develop a sequence of isomorphisms: we can map namely a closer approximation to D1 on a non-closer approximation to D2, or we can have the approximations to D1 and D2 closer but not the approximations to R and S, or we can have all closer but only some tighter, and so forth.

Approximation V isomorphism: let us be aware of the fact that in approximation IV both correlator and correspondence were exact relations. The great multiplicity of cases already encountered was due to the indetermination of D, R and S and not to that of correlator or correspondence.

It is now the place to mention that we can combine any form of approximation IV with approximations I, II or III. Here the situation becomes extremely complex and we can only stress this feature. Let it also be stressed that even in approximation I, II and III, the correlator and the correspondence had no indetermination domains. The two features we are thus liberalizing here, in V, are radically independent: we consider imprecise correlators and correspondences, and we consider moreover partial correlators and correspondences. It is perhaps best that we distinguish approximation V, alpha (subspecies: I, II and III), approximation V beta (subspecies: all subspecies of IV) and approximation gamma (both alpha and beta).

Approximation VI. Let it now be stated that in all the previous cases, even if we depart very strongly from the simple classical picture, we have been presupposing that a system can be described as a relational structure with given clear-cut relations and elements, the system having perhaps indetermination domains and the relations also, but these entities themselves being given and the wholes being built up out of these parts. It seems clear that this is a radical dependence on a type of logical atomism that geometry precisely tries to overcome through the construction of a geometry without points [6]). An algebra without elements seems as urgent a desideratum if we want to develop the model theory for natural systems. Let *n* systems be given. We do not presuppose the concept of element or of relation, but we define them with respect to these systems. An element is the smallest system that systems can have in common (as always in these topics, we presuppose the part-whole relationship and some of its properties), and a relation is a minimal system of sequences that systems can have in common. The definition of what is to be called element or relation should thus be a function of the intersection of systems. We cannot hope to develop here a complete theory for these approximate concepts. We hope to have shown that they exhibit interesting complexities.

Before we try to apply them in the domain of semantics proper, let us point out one direction in which the study of our approximate isomorphisms might profitably develop.

Alfred Tarski in his 'Contributions to the theory of Models' [7]) has defined the concepts of relational system, subsystems of relational systems, similarity of relational systems, homomorphism and isomorphism of relational systems, union and cardinal product of relational systems, and finally the concept of elementary or arithmetical classes of relational systems. The first task of the generalisation of formal semantics that seems necessary in the light of our study of empirical models in the empirical sciences, would be the application of some of the simpler forms of approximation defined above to the definition of these concepts. This is the more necessary because empirical models, as we have seen, are very often models through certain of their subsystems, and they are models of unions or cardinal products of other systems, all these concepts taken in some approximate sense. This task is not only necessary but possible. If a relational system is an arbitrary sequence consisting of a set, and of a series of finite relations of given rank, all defined on the elements of this set, then an approximate relational system is also a sequence where the first element is a set with an indetermination domain, where there follows a series of relations with indetermination domains, *nearly* all of them *nearly* everywhere defined on the set, having all of them ranks *within certain intervals*, and having *nearly* all of them finite rank.

The phrase 'nearly all' can either be replaced by some quantitative provision (giving the length of intervals, or the measure of sets), or by some qualitative provision (stating that intervals or sets are included as proper parts in certain others). Two relational systems will be approximately similar, if they are approximately of the same order (difference between order lying in a certain region), and if the relations can be correlated with each other so that the correlated ones have not too large differences of rank in too many cases (again, we do not indicate here the precise way in which we could write these phrases). A relational system is a subsystem of another one to which it is similar (one could replace this by 'approximately similar') if the set of the one is nearly completely included in the set of the other (the intersection of the first with the complement of the second being sufficiently small in metric or inclusion terms), and if the relations of the first are nearly all nearly identical to restrictions of the second to one of its subdomains. In terms of approximate subsystems, it is no longer true that SS(K) = S(K) as it is true for precise subdomains (though S(K) will be included in SS(K), where S is the set of subsystems of its argument and where K is any relational system). We do not have to repeat the definitions of isomorphism. There are as many generalisations of isomorphism as there are approximations mentioned. The approximate union of two precise relational systems (to be opposed to the precise union of two approximate relational systems and to the approximate union of two approximate relational systems) is the relational system that has, as set, an approximation to the union of the two sets, and, as relations, an approximation to the unions of the corresponding relations. The set of the cardinal product of two relational systems consists nearly completely of nearly all pairs of nearly this form (*a*-*b*), with *a* in R and *b* in S, and where the relations take, for *nearly* all their elements, pairs with *nearly* all their first members from R in the corresponding place and *nearly* all their second elements from S in the corresponding place. Here also a detailed investigation of the theorems about isomorphs (of different degrees of approximation) for unions and products (of different degrees of approximation) will yield necessary and interesting results.

Let us now however make a general remark: it is easy to define strict isomorphism for abstract relational systems given by their fundamental operations. For a group the relation to be preserved under mapping should be multiplication, for a ring addition and multiplication, for an ordered set the ordering relation, and so forth.

If we have however two physical systems, or two languages, how are we going to pick out the relationship that is to be preserved under mapping?

It may be shown that even for physical systems a natural isomorphism concept can be defined. It has been recently most clearly explained by Ross Ashby, in the paragraphs on models occurring in his recent book. [8])

Let us however stress that Hertz's definition of dynamical similarity is close to Ashby's [9]). The relation that is required to remain invariant under the mapping is here the relation of 'going from one state into another under the influence of an input'. We characterize a physical system by an input-output matrix, or by a matrix giving for every input and present state the following output, or giving for every input and every state the next state. All these possibilities are open, and mostly equivalent. On p. 98 Ashby then states 'the canonical representations of two machines are isomorphic if a one-one transformation of the states of one machine into those of the other can convert the one representation into the other'. Only for strictly deterministic systems, however, did Ashby provide his definition (i.e. for systems we know completely, and we are interested in for their complete behaviour). We have seen repeatedly in our second section that only when systems are partly known is modelling needed. We shall thus have to adopt definitions of physical isomorphism for systems whose input-output (or state, or input-state, or output-state) matrix we cannot completely write down. We shall have to consider probabilistic systems (for which I can only give the probability for a given input to have a given output), or even weaker ones (for which I can only sometimes say that one output is more probable than another for given input and state). Ashby is aware that he must liberalize the relation between prototype and model and cannot identify it with the relation between isomorphs. However the only way he succeeds in doing so is to say that a model M is such that one of its homomorphs is isomorphic to one homomorph of the prototype. This liberalization still obliges us to know the complete deterministic state matrix of both model and prototype, and thus does not yield the desired result.

It is clear that we can always represent a relation by a matrix and that the approximations to isomorphism we have been considering should find their adequate translation in conditions on the reciprocal relationships of matrices. These reciprocal relationships will be the real translation of the modelling relationship between physical or biological systems, as it occurs in physics and biology. The contribution of Ashby consists however in having shown that the state matrix is the adequate tool to define isomorphism between natural systems, even if the liberalization he proposes for the classical relationship is not one we can adopt. His approach allows us to assert that our approximate isomorphisms will easily find application in this domain.

B. *Semantic models*

The concepts of truth, designation, satisfaction and definition are closely related to each other and the search for a liberalized version of one of them will be reducible to the search for a liberalized version of some others in this series. Let us mention in this context that Herbert Simon, in his recent paper 'Definable Terms and Primitives in Axiom Systems' [10]), proposes two generalisations of the concept of definability that correspond immediately to generalisations of the concept of satisfiability. A language L, according to Simon, has a generic definition for an individual *a*, if in L we can write down the definition for a class to which *a* belongs (i.e. a necessary condition for *a*). A language has an approximate definition for *a* if in L we can write down a necessary and sufficient condition for any *x* to be *a*, except if *x* belongs to a certain set of measure zero.
A domain generically satisfies a certain statement or set of statements if it satisfies certain statements belonging to the same class as these, or if some model belonging to a class K with the present one, satisfies the set of statements.
A domain satisfies approximately a set of statements if every part of it, except a subset of measure zero, satisfies this set of statements.
It is clear that our transposition of Simon's general definitions for definability towards definitions of satisfiability is perfectly possible. It is however difficult to believe that the two concepts Simon introduces are the only or most fruitful ones: generic definability or satisfiability is extremely weak and completely depends on the criterion of class membership chosen; approximate definability or satisfiability strongly depends upon the existence of a measure function, and on infinity and continuity considerations (all denumerable sets having measure zero).
We feel encouraged by this attempt in our search for an enlarged and liberalized relation of satisfaction, but we do not think the main task is already done.
In 'Logic, Semantics and Metamathematics' (p. 416), A. Tarski [11]) defines the concept of model. Let in language L to every extra logical constant correspond a correlated variable in L, in such a way that every sentence becomes a sentential function if the constants are replaced by the variables. An arbitrary sequence of objects satisfying these functions will be said to be a model or a realisation of that class of sentences. The semantic model concept is thus both akin to and very different from the ideal

of modelling relationship as isomorphism. The prototype-model relationship is here by definition anti-reflexive, anti-symmetrical, but transitive; while in its algebraic version it is reflexive, symmetrical and transitive. But whenever *n* systems are models of the same language L, there exist between these systems, or between parts of them, relations of isomorphism. This implies that to the generalisations for isomorphism we have introduced, there must correspond generalisations of the satisfaction relation. The main problem of this section is: if we claim that among *n* domains some specific approximate isomorphism exists, is there (a) a language they then all strictly satisfy or (b) is there a language they all approximately satisfy? The solution of these two problems will not be given here; we only ask the question. But a step towards the solution will certainly be taken if we define some of the natural liberalized versions of satisfaction itself. Let us give the following definitions:

A set factually satisfies a sentence *p* of a language L if and only if the variables of *p* range over the set S, the predicates of *p* over subsets of S, the logical constants are interpreted as usual, and the sentence becomes true for S.

We shall say that approximate satisfaction can be defined with reference to assignments for individual variables, for predicate variables, for logical constants, or for truth.

We shall number these types of approximate semantic satisfaction as Approximate satisfaction 1, 2, 3 or 4.

A set appr. 1 satisfies a sentence if for variables ranging over some subset or superset of S, *not too distinct* from S, all other properties remain identical.

The closeness of the approximation will again be measured by the inclusion relations of the Inters (Range Variable, Complement S).

A set appr. 2 satisfies a sentence *p* if the predicates of *p* range over a class of subsets of S plus or minus certain elements, and all other properties hold true.

In approximation 3, logical constants could be otherwise defined than usual (through the rules of propositional calculus and functional logic). Here, in order to define what is the ordering principle of the approximation, we should be able to define what is a closer approximation either to negation or conjuction itself or to the classical interpretation of these constants.

Finally approximation 4 replaces true by an approximation to true (the one that comes immediately to the mind is: probable, but we are by no means forced to select this type of approximation).

It is obvious that if we give ourselves weaker conditions and preserve the same conclusion, we then obtain a strengthened version of the satisfaction requirements.

In this respect, however, it is, we think, interesting to refer to certain approximate but strengthened forms of satisfiability.

Let us say that a domain constructively satisfies a language L if for every sentence of L we have a procedure that allows us to construct the domain in question and the various individuals and classes needed for the verification of the truth of the sentences. This procedure for geometrical concepts is even expressed in L when everything is so defined that the postulates given realise a method of measurement for these concepts. This constructive feature may now only approximately exist, either as a procedure given in most but not in all cases, or as a procedure yielding close but not completely exact correspondence. This we should like to call the approximate constructive satisfaction relation.

Let us say that a domain necessarily satisfies a sentence if many possible assignments within this domain, or many possible extensions of this domain, or many possible other domains, satisfy this sentence. The degree of necessity could easily be ordered. We can then speak about necessarily approximate interpretations and approximately necessary interpretations; the approximate interpretation can hold perhaps in many other selections or domains; or in most but not all of a class of chosen domains.

The relationship between formal languages and domains in which they have models must in the empirical sciences necessarily be guided by two considerations that are by no means as important in the formal sciences: (a) the relationship between the language and the domain must be closer because they are in a sense produced through and for each other; (b) extensions of formalisms and models must necessarily be considered because everything introduced is introduced to make progress in the description of the objects studied. Therefore we should say that the formalisation of the concept of *approximate constructive necessary satisfaction* is the main task of the semantic study of models in the empirical sciences. Here however, once more, we should stress all the difficulties of the undertaking: approximate modalities and approximate recursiveness should be the basic tools of this enterprise, basic tools that should then be applied to the model-prototype relationship.

We meet here another major problem for our generalised semantics: if we decrease the requirements a system has to satisfy in order to be a model for a language (as we are constantly doing when we apply approximate concepts), and if we apply simultaneously strengthened model concepts (as we do when we speak about recursive and necessary representations), can we in some sense compensate our first loss of properties by a corresponding gain, so that our results are either identical or at least equal in power of deduction? This is perhaps the deepest problem we should try to solve in building up these approximate systems of semantics.

Let us give some more details about liberalizations of semantic concepts, details whose function it is to apply the notions of our section III A to the most popular model definition. Usually, a model is defined in two steps, the first of which defines what is called a semi-model and the second of which gives the full model.

The domain D is circumscribed on the one hand, the language L on the other hand. Simple expressions are defined in L, classes of elements are defined in D. Rules are then given to assign to each non-complete expression in the language an element, a class or a sequence of elements in the domain (these elements, classes or sequences may belong to very many different categories, first defined). It is then shown that every term has an assignment. This assignment or semi-model is moreover a model if all asserted sentences in that language are assigned in their category to one definite sub-category, in which nothing else but these asserted sentences is assigned the truth. The recursive building-up of higher-order categories and of higher-order complex sentences is thus essential for the definition of model. This being the general description of what it means to define a model (as stated for instance by Kemeny and Tarski[12])), it is now clear what will be the dimensions along which approximations will have to be defined:

(1) approximations to the language on the one hand, the domain on the other hand; (2) approximations to the category of simple signs on the one hand, to elements of the domain on the other hand; (3) approximations to the subdivision of the domain of simple signs in L, of the domain of simple elements in D; (4) approximations in language and domain to the several complexity-creating operations; (5) approximations to the vertical

division in expression-classes or element-categories; (6) approximations to the various assignment rules that have to be combined with all the other types of approximation. To be brief, we could say that all the earlier categories of an algebraical nature can here be applied but that *we must apply them to two hierarchised structures built up out of a given basis by a few iterated operations.* The semantic approximation should thus be approximation to a correspondence between two hierarchies. This is the only new and important feature that occurs. But along all the mentioned dimensions of approximation, all forms of approximation distinguished have to be applied, and moreover they have to be combined (the major problem being: How is the approximation form and degree at a higher or later stage dependent on the approximation level and degree at a lower level and stage?).

We should however be aware of the fact that it is difficult to discuss in general liberalizations of the satisfaction or model concept. Only for specific languages can this concept be defined, because the sentences stating what types of entities satisfy what simple sentences are purely relative to the language in which we find ourselves. We should thus in fact select a given language and produce a liberalized version for the satisfaction concept there. This cannot be our aim here, but the broad outlines of such an undertaking have, we think, sufficiently been sketched.

To summarize our results: the more complex an entity, the more difficult it is to define an approximation to that entity, because so many different dimensions exist along which the approximation should be defined. A language is such a complex structure, and defining approximate semantic isomorphism means defining an approximate relation of correspondence between approximations to a language and approximations to a domain (set).

We see that the difficulties we encounter could in some sense depend upon the notion: approximation to a language.

This is the central concept of the last part of this section.

C. *Syntactical models*

Here we are going to study the relation between model and prototype as a relation between two languages. We consider as languages the different sciences, and this forces us immediately to consider certain generalisations with respect to formal calculi. For most scientific systems, the distinction between axioms and rules of inference, for instance, is not clear. The series of definitions is not closed and it is not a mechanical procedure to know if a given sentence belongs to the system or not. Most scientific systems do not have a clear distinction between logical and descriptive constants. In fact a science seems to be written in a mixture of everyday language describing experiments, mathematical equations describing calculations and semi-formalised deduction if the science is sufficiently advanced to contain a theoretical part.

We neglect all semantic and pragmatical features, because we consider here only the syntactical aspects.

In opposition to this very complex mixture that is a science (when this science is taken as it is and not transformed for philosophical or logical reasons into something else), what is a calculus? A calculus is a sequence of the following entities: a set of signs, a set of sequences of these signs that are well-formed formulae, a set of well-formed formulae that are theorems, and a set of axioms (or, equivalently, a set of sequences of well-formed formulae that are rules of inference). The sequence (S, F, T, A) is thus a calculus. We can assert certain evident relations between the different elements of this four-term sequence and one of the most interesting properties is that S, F, A ought to be recursive and T recursively enumerable.

The facts just mentioned about an empirical science seem to indicate that there the elements of this sequence are not recursive, that each of them has indetermination ranges, and that moreover the four ingredients of the calculus have indetermination regions in common. This leads us to believe that if we start with the general concept of calculus, and if we define various types and degrees of approximation to calculi (and mixtures of calculi), we shall probably have among them the specific features of our empirical sciences.

A language L will be called a closer approximation to a calculus than a language K if and only if (a) there are more signs in L that are indisputably simple and belonging to the calculi than in K where either fewer signs are indisputably simple or fewer simple ones are decidably elements of the calculus; (b) there are more sequences in L than in K that completely consist of signs of S and that are moreover indisputably well-formed (against cases in which sequences have as elements border-line signs or are themselves on the borderline of well-formedness); (c) there are more sequences that completely fall into the kernel of the set F and that are in the kernel of T; (d) there are more sequences indisputably consisting of members of S, indisputably in F and T, and members of A. This defines the general idea of approximation of a language to a calculus in general. When now is a language an approximation to a given calculus in particular? We think that we could here give a fairly general answer. In fact, a calculus is dependent upon a classification of signs, of sequences of signs, of sequences of such sequences and sub-classifications of these. It is an ordered sequence having as elements classes chosen from such classifications. The principle of the order exhibited in this sequence is given by the properties: T lies inside F, A lies inside T, all parts of members of F are members of S, and none of the inverse characteristics hold. We can say that a classification is an approximation to another classification the more classes of the first coincide more completely with classes of the second.

A hierarchy of classifications is the more an approximation to another hierarchy, the more each level corresponds to the corresponding level, and the more the relations between each pair of successive levels mirror the relations between the other pairs of corresponding levels. The degree of approximation of one hierarchy to another depending thus on at least two factors (and on the definition of corresponding level), the same degree of approximation could correspond to very different situations. Presumably proper weights should be chosen to determine the importance of each factor in the determination of the closeness of approximation to a given calculus. A calculus being essentially a selection from a given hierarchy of classification, this trend of thinking can lead to a definition of the ordinal degree of approximation of one calculus to another. The limit of a class with respect to a relation (in *Principia Mathematica*) is an element such that it stands in the converse of R to every element of that class and such that for every element in the class there is another element in it such that the first has the relation R to the second. If we now take as elements classes and as the relation a relation of inclusion we can define the limit of a series of classes; and, as a consequence, also the limit of classifications and of hierarchies of classifications. Once this is done nothing can prevent us from defining limits of languages. We certainly could solve the problem of ordering approximations to languages without defining limits for sequences of languages, but it is certain that if we could define such limits, our ordination difficulty would most neatly be solved. We must however limit ourselves here, more than ever, to tentative non-formal remarks.

Let us repeat that we could define neighbourhoods for languages and calculi in function of the neighbourhoods eventually defined for S, F, T and A (the classical neighbourhood-axioms could presumably easily be satisfied).

The syntactical model concept is defined in Tarski's book *Undecidable Theories* essentially for 'systems in standard formalisation'. These systems are calculi in the sense we have just discussed with very careful sub-classifications of the series of signs, and of sequences of signs, and of the series of axioms. We can neglect the particular nature of these sub-classifications as the general problem of approximation remains the same for all these refined versions.

Having thus understood how we can relate systems in standard formalisation to languages in science, we can now come to our main topic, the study of the syntactical interpretation relation as defined among standard formalisms, and its approximations; this will be an introduction to the study of generalisations of this syntactical relation as defined among approximations to standard systems.

Tarski tells us, on pp. 20-22, for systems in standard formalisation, that a system is interpretable [13]) in another one if we find in this other system a series of definitions for the basic terms of the first that give to these basic terms their usual properties.

Tarski, using his distinction between logical and non-logical constants, applies this first to the case where only for non-logical constants are definitions contemplated, and where among the descriptive terms, only constants are to be interpreted. Later he admits, in a footnote on p. 22, that logical constants may also be interpreted, and in his method for the relativisation of quantifiers he in fact also reinterprets variables. We shall come to this later.

How might we consider weakened forms of these situations? We could envisage only partial definitions, or definitions of only part of the basic terms, or definitions giving to these basic terms only part of their earlier properties; or definitions that are not partial but multiple and probabilistic (every sign receiving a sequence of definitions with different cardinal or ordinal probabilities). These three directions of generalisation still presuppose that the system to be interpreted and the interpreting system are both in standard formalisation! Only the interpretation relation is weakened.

Let T1 and T2 not have any non-logical constants in common, then T2 is interpretable in T1, if and only if there is a theory T and a set D, satisfying the following conditions: (a) T is a common extension of T1 and T2, every constant of T being a constant either of T1 or of T2; (b) D is a recursive set of sentences, valid in T, and possible definitions in T1 of non-logical constants of T2; (c) every non-logical constant of T2 occurs in at least and at most one sentence of D; (d) all valid sentences of T are either those of T1, or of D [13]). To understand the force of the definition it is necessary to understand that a theory is an extension of another one if every sentence of this other one that is valid (we should try to find some syntactical equivalent for this term) is also valid in the first.

Tarski himself weakens his definition in one direction, but let us first stress the following points:

– T might not be a common extension of T1 and T2; in other words: in order for the interpretation to be possible, certain properties that hold for T1 would cease to hold;

– the set D might not be recursive; its contents might not be possible definitions for the constants of T2 but only partial ones, or only properties, or only probabilistic sentences;

– the non-logical constants of T2 might occur in more than one sentence of D (over-determination) and not all of them might occur;

– there might be valid sentences of T neither in T1 nor in D (i.e. certain properties of the defined signs are no longer derivable in the interpretation and must be added independently).

These possibilities all go in the direction of the weakenings we contemplated informally above.

Tarski's own 'weak interpretability' goes in the direction of our last possibility (extension is preserved, all constants must be the same, but new truths about the same topics might be needed). Our last possibility is wider than his weak interpretability because it considers interpretation in a super-theory, not having necessarily the same constants.

If we now cease to consider only formalisms in standard formalisation and if we try to introduce approximations to the standard formalisation,

we encounter two very divergent cases that have to be studied: (a) can we adapt Tarski's 'strict' definition of interpretation to the case of approximations to standard formalisations, and
(b) can we adapt the weakened forms of it we just suggested to approximations to standard formalisations?
The first case covers two different situations; it may be that both the systems T1 and T2 are approximations to systems in standard formalisation, or that one of them is a standard formalised system (more generally: the mode and degree of formalisation in both cases may be the same or different).
If, for instance, instead of knowing or not knowing that signs belong to T1 or T2, we have a continuous spectrum of probabilities and know that the probability that they are in T1 belongs to a given probability interval, we might paraphrase the first condition Tarski puts forward for interpretation to exist, as follows: if a given sign lies in a probability interval *i*, regarding T, then it lies in the same probability interval regarding either T1 or T2. (We could loosen this up by saying that there is at least one among the two languages such that the probability interval of this sign in this language is not too far away from the corresponding interval in T). This probability might be ordinal.
Such a scheme tries to keep between two not completely formalised systems the same strict relationship that was asked for in the case of two completely formalised systems.
Let us try to look for a similar transcription of the first demand for another type of approximation: let it be given that a lump of signs or sign sequences belongs to a language, there being no method available to dissect this union, the elements of which remain undiscriminated. This entails the presence of a union of sentences equally indistinguishable and speaking about the given signs. We could now, by analogy with the earlier procedure, say that the lumps in T should be such that they coincide either with some in T1 or with some in T2 (or we could say that every element present in a lump of T should be also present in some lump of T1 or of T2). It is obvious that we here try to apply to the concept 'formal system' in the first place and the concept 'interpretation' in the second place, some of the concepts of approximate geometry introduced by Hjelmslev and applied already by Menger to the calculus of relations.
This analogy suggests that, exactly as in the geometry of solids the point

is defined by a series of approximations, we should develop a syntax and semantics of approximately defined systems that might take over some of the techniques of this geometry of solids, the problem that has been tackled in both cases being the same: starting from a formally clear situation, approach reality by generalisation that makes lesser demands on powers of discrimination.

The situation we have encountered in our attempts to preserve the strict modelling relationship for not-strictly-defined systems is a situation we have by no means exhausted (we should stress that for each of the rules of interpretation and for each of the modes of approximation new difficulties arise). To see what types of difficulties might present themselves, let us look at a closely related case: suppose we admit logical constants themselves to be interpreted. How then to interpret the demand that the logical relations of the interpreted symbol of descriptive nature should be the same as those of the initial symbol?

So also the interpretation rule for signs must be formulated so as not to contrast with the interpretation rule for sequences and so forth.

Let an approximate interpretation be called adapted to the mode of approximation of the languages that are model and prototype, if the dimensions along which the interpretation approaches a formally complete interpretation are the same as those along which both languages approach the status of formal calculi. An interpretation can certainly be unilaterally adapted, because the mode of approach of model and prototype is not necessarily the same. An interpretation can certainly be of different degrees of inadaptation. For instance, let model and prototype be close to formal calculi except for *n* signs, whose belonging to the system is uncertain. An interpretation is then adapted if it is a formal interpretation except for *n* rules of assignment for simple signs, that remain equally uncertain. If the form of uncertainty is specified (as a degree or through a disjunction) we can even require closer analogy. Let it be said however that the signs for which the assignation rules are undetermined are not necessarily those for which uncertainty of belonging exists. Even for approximate interpretations that are adapted to their terms such a multiplicity subsists. How intricate will be the complexity of approximate interpretations not adapted to the mode of approximation of their terms!

We want now to complete our survey of modes of syntactical interpretation by the study of Tarski's relative interpretation.

Given a theory and a one-place predicate P we construct an interpretation by relativisation with respect to P if we replace all sentences quantified and including x, by sentences quantified and having the hypothesis that their x's have the predicate P. Relativisation with respect to a predicate of the domain of individuals should perhaps be more generally defined as: relativisation of a domain of entities (propositions, predicates themselves, relations), with respect to an element of a domain of entities that they can be applied to or that can be applied to them by admissible operations (in a propositional calculus we might replace every p by a clause ($q\mathrm{I}p$) for instance; similar changes might occur in higher functional calculus).

It begins to be rather tiresome when we stress that the relativisation clause might be imposed only on certain predicates, or might be imposed diversely with diverse probabilities, or might concern a non-unique predicate but the union of a sequence of predicates.

The relationship between interpretation by relativisation in the strict case and interpretation by definition is not yet completely clear; it should not astonish us that *a fortiori* the relationship between approximations to these two types of interpretation on the syntactical level is a topic still awaiting investigation.

If a science is really something rather different from a formal system, it should meet not only weaker demands, but also stronger demands. Here our problem is much more undetermined than elsewhere: we know from long experience how to generalise certain conditions; but it is not clear how we should strenghten them.

Certainly, from the syntactical point of view, a science is not only any axiomatisable or unaxiomatisable formal system but has some stronger type of unity. The axioms should not be unrelated. They should in some sense be 'about the same topic'; moreover the set of theorems should present, first, some type of symmetry, second, some type of mutual involvement. How formally to represent this type of symmetry, unity or involvement as strengthened versions of the requirements for 'systems in standard formalisation' is far from clear. Moreover, it is quite true that the usual models, even from a syntactical point of view, are not given by arbitrary rules of definition for the constants of T2 in T1. The definitions should delimit a domain of objects that in T1 itself is necessary as a separate domain, having sufficient distinctness and standing out against other parts of T1 having similar distinctness; to derive from the rules in D the properties of the defined terms, the derivations should also have some sufficient unity, showing that the defining characteristics are in some sense grasping the nature of the constants in question. How to define the naturalness of this inference, and thus of the definition, is however difficult to see.

Heuristically we could follow two directions: we could ask what are the strengthenings of the definitions that could in some sense compensate for the weakenings we proposed earlier? And we could ask: if we consider the given definitions as obtained by generalisation from other more demanding ones in the same way as we obtained our generalisations, what could be our starting points?

Let us be satisfied that here too there is a purely formal way of reaching the more restricted relationships that we need for reasons of adequacy. It is here that we must abandon the topic of approximate model building for purely syntactical systems. We think that of the three dimensions we have explored each shows its own direction of generalisation: generalising the isomorphism of arbitrary relational structures, or the satisfying of a language by a domain, or the translatability of a language into another, are three fundamentally different operations.

IV. ATTEMPT TOWARDS REUNIFICATION

The results of the study presented in this paper show that we cannot hope to give one unique structural definition for models in the empirical sciences. If a unification is still to be possible, we should go back to our starting point: the function of models. If we can give a strict and formal definition for the function of a model, we can – this is our final impression – hope to reach on new grounds a general description of our multiform concept. It is here that our attention should turn towards formal pragmatics [14]). A subject uses a language to reach certain aims. If this notion can be formalised, it is also possible to formalise the notion that a subject uses a language to obtain information about another one, or uses a physical system to obtain information about another one. This will be our final and most general hint towards the definition of model: *any subject using a system A that is neither directly nor indirectly interacting with a system B to obtain information about the system B*, is using A as a model for B. The definition of 'using', 'purpose' and 'information about' are problems formal pragmatics is already beginning to tackle. While we do not think that this type of definition of the model concept is very fruitful (the syntactical, algebraic and semantic study of the various special model concepts seem to us immensely more fruitful), we are convinced at least that a general definition along these lines is possible, adequate and formal.

Universities of Ghent and Brussels

REFERENCES

1. R. C. Braithwaite: The Nature of theoretical Concepts and the Role of Models in an advanced Science *(Revue Internationale de Philosophie, 1954, fasc. 1–2)*.
2. M. B. Hesse: Models in Physics *(British Journal for the Philosophy of Science, 1954, pp. 198–214)*.
   E. H. Hutten: The role of Models in Physics *(BJPS, 1954, pp. 284–301)*.
3. Rudolf Carnap: *Introduction to Semantics*, Harvard University Press, 1946.
4. John G. Kemeny: A New Approach to Semantics, Part I *(Journal for Symbolic Logic, 1956)*.
5. For information about dimensional analysis, consult N. R. Campbell: *Foundations of Science* (ch. XIV and XV) and P. W. Bridgman: *Dimensional Analysis*.
6. For information about approximate geometry, consult Karl Menger: *Rice Institute Pamphlets*; *1940*: Geometry without points, and the same author: Géométrie Générale, *Mémorial des Sciences mathématiques, 1950*.
7. A. Tarski: Contributions to the Theory of Models *(Proceedings Kon. Ned. Ak. Wet., 1954)*.
8. A. Ross Ashby: *An Introduction to Cybernetics* (Wiley 1957) pp. 102–109, 'homomorphic machines'.
9. Paul Hertz; *Principles of Mechanics* (Dover Publ., par. 418–428 Dynamic models).
10. Herbert A. Simon: Definable Terms and Primitives in Axiom Systems *(The Axiomatic Method in Geometry and Physics, pp. 443–453)*.
11. Alfred Tarski: *Logic, Semantics and Metamathematics* (Oxford, 1956).
12. See Kemeny, op. cit. and Tarski, op. cit.
13. Tarski, Mostowski and Robinson: *Undecidable Theories* (Amsterdam, 1952).
14. Richard Martin: *Toward a Systematic Pragmatics* (Amsterdam, 1959).

RICHARD C. ATKINSON

THE USE OF MODELS IN EXPERIMENTAL PSYCHOLOGY [1]

In this paper I shall not be concerned with a formal analysis of the function of models in psychology. The problem has been considered on many occasions by both psychologists and philosophers, and I am not inclined to add to the voluminous literature in this area. Instead, I shall describe a fairly simple model of behavior and illustrate the method of application to a complex problem in decision making. By examination of this particular case we will be able to indicate the role of mathematical models in determining programs of psychological research and specifying the types of empirical observations to be made.

The case to be examined deals with the psychology of learning. There are three basic concepts in this area which play a central role in both theoretical and experimental work; they are the concepts of stimulus, response, and reinforcement. The stimulus is conceived as an environmental event; a response is an act or movement made by the organism exposed to the stimulating situation; and a reinforcement is any event (experimenter or subject controlled) which gives rise to an increment or decrement in the likelihood that a stimulus will elicit a particular response. One of the principal problems of learning psychologists is to specify the relations between stimuli and various response measures as a function of reinforcement schedules. A major contribution of early behavioristic scientists like Pavlov, Thorndike, Bechterev, Watson, and Guthrie was the development of these concepts in a scientific sense, clearly disentangling them from notions of common sense and of earlier philosophical psychology.

The introduction of mathematical formulations as a tool in the analysis of learning has been of fairly recent origin. Some of the earlier work by Schükarew, Robertson, Thrustone, Woodrow and others was concerned with finding an analytic function which was to provide a universal description of learning. Many suggestions for the appropriate function

[1]) This work was supported by the Office of Naval Research under Task NR 170–282.

were made and there was much debate as to which was correct. This search for the universal learning function began with *ad hoc* proposals and curve fitting to experimental data. Gradually, however, these endeavors gave way to the more constructive undertaking of providing systematic formulations of the elementary events underlying the learning process.

One of the most noteworthy programs in constructing a quantitative theory of learning was that of Clark L. Hull. In his *Principles of Behavior* he presents a set of postulates designed to encompass the major aspects of learning. From these postulates deductions were made which initiated a great deal of empirical research. Unfortunately, Hull's work, and that of his contemporaries like Tolman and Lewin, did not lead to a theory that was mathematically viable. That is, in Hull's system it is possible to make only a very limited number of derivations leading to new quantitative predictions of behavior.

However, the work of Hull, Tolman, Lewin and others emphasized the importance of rigorous theory construction in psychology and set the stage for recent developments in mathematical learning theory. The work by Estes [1950], Bush and Mosteller [1950] and Estes and Burke [1953] initiated these new developments and represented analyses of learning which led to mathematically tractable systems. The work of these investigators and subsequent work of Luce, Suppes, Restle, Audley, and many others has resulted in systematic formulations of learning which have the same sort of feel about them that theories in physics have. Nontrivial quantitative predictions can be made – not only about the gross phenomena of learning but also with regard to the fine structure of the data. Once appropriate identification of theoretical terms has been made it is usually clear how to derive predictions about responses in a manner that is not *ad hoc* and is mathematically exact.

To illustrate some of these points I would like to present a particular set of axioms for describing learning. Only those axioms will be presented that are necessary for the analysis of the experiment to be considered in this paper. The reader interested in a more comprehensive formulation is referred to Suppes and Atkinson [1960].

The experimental situation consists of a sequence of discrete trials. There are K response alternatives, denoted $A_i (i = 1, \ldots, K)$. On each trial of the experiment two or more alternatives are made available to the subject, and he is required to select one of the available responses. Once his response has been made the subject wins or loses a fixed amount of money. The subject's task is to win as frequently as possible. There are many aspects of the situation that can be manipulated by the experimenter but in this paper we will consider only the following variables: (1) the strategy by which the experimenter makes available certain subsets of responses on any trial of the experiment, (2) the schedule by which the experimenter determines whether the occurrence of a particular response by the subject leads to a win or loss and (3) the amount of money won or lost on each trial. The role of the model in this situation is to provide an explicit and detailed account of a subject's responses over trials of the experiment. One reason for investigating this particular experimental problem is that it is a prototype of many decision making situations in the real world. If behavior can be predicted with accuracy in our laboratory situation, then we shall have substantially increased our understanding of decision processes in general.

The model we shall consider assumes that (1) associated with each response alternative there is a tendency to approach or avoid that alternative and (2) the response which is finally made on a trial depends on the observing or orienting behavior of the subject in the pre-decision period of the trial. The basic notions underlying the model are similar to those presented by Bower [1959], Estes [1960] and Audley [1960].

The axioms will be formulated verbally. It is not difficult to state them in a mathematically exact form, but for our purposes this will not be necessary.

A1. *On every trial each response has an approach-avoidance value* (AAV) *of* 1 *or* 0.

A2. *At the start of each trial the subject randomly observes one of the available responses.*

A3. *If the* AAV *for a particular available response is* 1 *and the response is observed, then that response will be made. If the* AAV *is* 0 *and the response is observed, then the subject will randomly reorient and observe one of the other available response alternatives.*

A4. *If all available responses have been observed on a trial and no response has been selected* (*i.e. the case where all available responses have* AAV's *of* 0), *then the subject terminates the trial by randomly selecting one of the available responses.*

A5. *If a response is selected on a trial and followed by a win, then with probability ρ' its* AAV *becomes* 1 *and with probability* $1 - \rho'$ *its value remains unchanged. If a response is selected and followed by a loss, then with probability ρ'' its* AAV *becomes* 0 *and with probability* $1 - \rho''$ *its value remains unchanged.*[1]) *The reinforcement parameters ρ' and ρ'' are independent of the trial number and the preceding pattern of events.*
A6. *The* AAV *associated with a response not selected on a given trial does not change on that trial.*
Several experiments have been conducted to test the adequacy of these axioms, but we shall restrict ourselves to one reported by Suppes and Atkinson [1960]. Subjects were run for 360 trials and on every trial they won or lost a fixed amount of money. There were four responses (A_1, A_2, A_3 and A_4) and on each trial exactly two of these responses were made available to the subject; the six possible response pairs occurred with equiprobability. On each trial the subject was required to select between the two available responses but was given no other information. A win or loss on a trial depended on the response selected. If A_i was available and chosen, then with probability ξ_i the subject won and with probability $1 - \xi_i$ he lost. The values used were as follows: $\xi_1 = .2$, $\xi_2 = .4$, $\xi_3 = .6$ and $\xi_4 = .8$.[2]) Thus, if a subject is to maximize his probability of a win, he should choose A_4 whenever it is available, A_3 when it is available and A_4 is not available, and finally A_2 if neither A_3 nor A_4 is available.
For mathematical analysis in the remainder of this paper it will be useful to introduce the following notation:
$D_n^{(ij)}$ = the experimenter controlled event of making response pair (A_iA_j) available on trial n of the experiment ($i \neq j$).
$A_{i,\,n}$ = selection by the subject of response A_i on trial n.
W_n = a win on trial n.
W_n' = a loss on trial n.

In terms of the axioms we define the subject-state on any trial of the experiment by an ordered four-tuple $< ijkl >$ where $i, j, k, l = 1$ or 0. The first entry denotes the AAV assigned to response A_1, the second the

[1]) We presume that ρ' and ρ'' are monotone increasing functions of the amount of money that can be won or lost on a trial.
[2]) A quite different model for this situation has been proposed by Suppes (1959).

value for A_2, and so on. From the axioms it can be shown that, for our particular experimental procedure, the sequence of random variables which take the subject-states as values is a Markov chain. This means, among other things, that a transition matrix $P = [p_{ij}]$ may be constructed where p_{ij} is the probability of being in subject-state j on trial $n + 1$ given subject-state i on trial n. The learning process is completely characterized by these transition probabilities and the initial probability distribution on the states.

To illustrate the application of our axioms, we will derive one row of the transition matrix. In making such a derivation it is convenient to represent the various possible occurrences on a trial by a tree. Assume that we are in state $< 1001 >$ on trial n, then the appropriate tree is given in Figure 1. As indicated on the top branch, when the response pair (A_1A_2) or (A_1A_3) is made available (with probability $\frac{1}{3}$) the subject will select A_1, because the AAV is 1 for A_1 and 0 for both A_2 and A_3. If a win follows the occurrence of the A_1 response (with probability ξ_1) no change in the AAV's occurs; however, if a loss terminates the trial (with probability $1 - \xi_1$) then with probability ρ'' the AAV associated with A_1 becomes 0 and the new subject-state is $< 0001 >$. When the response pair (A_1A_4) is presented (with probability $\frac{1}{6}$) the subject selects the first response observed, since the AAV for both A_1 and A_4 is 1. When response pair (A_2A_3) is presented both available responses have AAV's of 0, and by Axiom 4 the subject randomly selects either A_2 or A_3. The other paths of the tree are obtained in similar fashion.

Each path on the tree from a beginning point to a terminal point represents a possible outcome on a given trial. The probability of each path is obtained by multiplication of conditional probabilities. Thus in Figure 1 two paths lead from $< 1001 >$ to $< 0001 >$ and the corresponding transition probability is $\frac{1}{3}(1 - \xi_1)\rho'' + \frac{1}{6}\frac{1}{2}(1 - \xi_1)\rho''$.

Construction of the other trees yields a transition matrix for a sixteen state Markov chain. Certain states in the chain will be transient if some of the probabilities ξ_i are 0 or 1.[1] However, in the experiment to be discussed this condition did not hold; therefore, our comments will be confined to the case where the ξ_i are different from 0 and 1. For this case, state $< 1111 >$ is transient (in fact, the probability of re-entering

[1] A state of a Markov chain is transient if the probability of ever returning to it is less than 1.

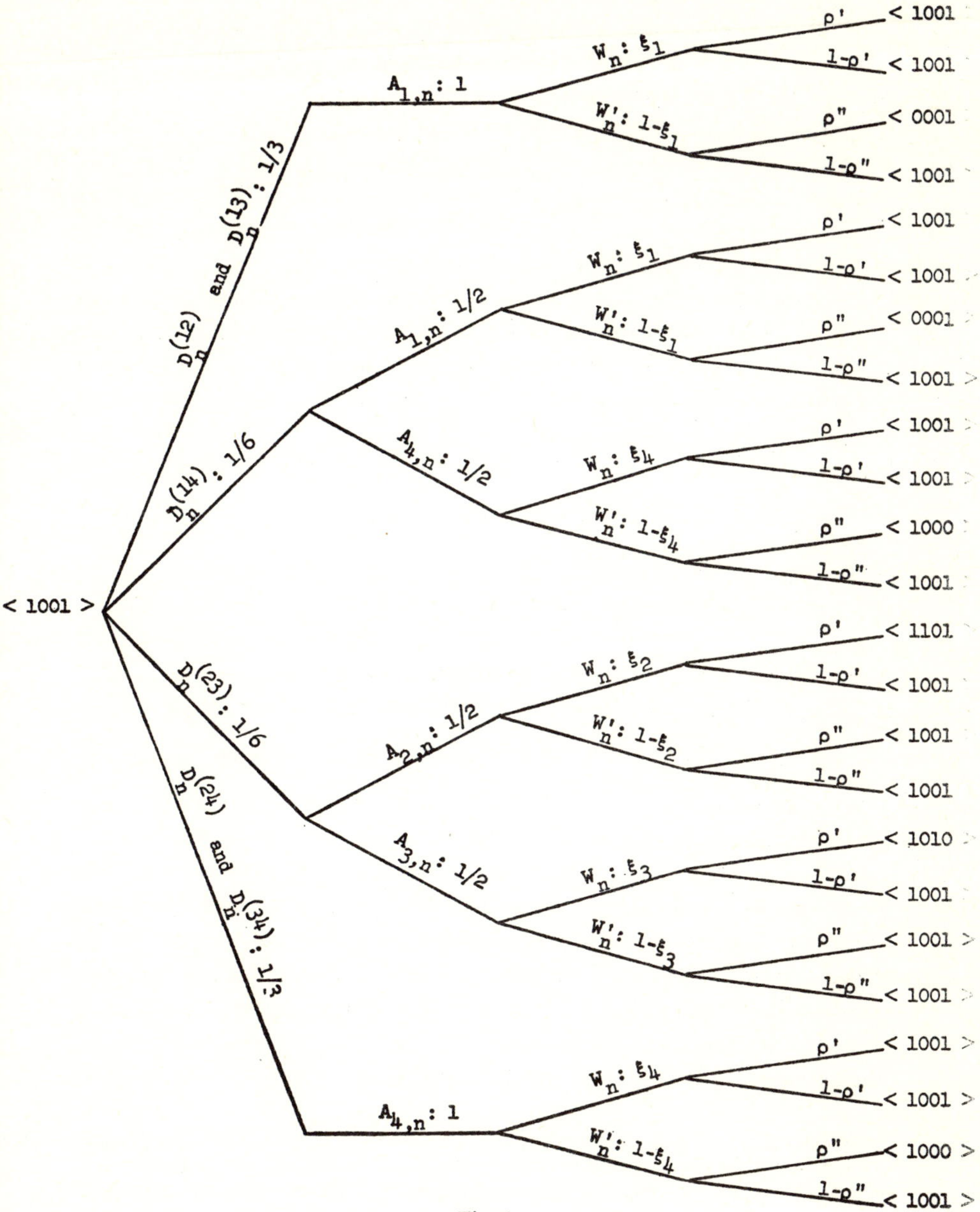
< 1001 >
D(12)n and D(13)n: 1/3
D(14)n: 1/6
D(23)n: 1/6
D(24)n and D(34)n: 1/3
A1,n: 1
A1,n: 1/2
A4,n: 1/2
A2,n: 1/2
A3,n: 1/2
A4,n: 1
Wn: ξ1
W'n: 1-ξ1
Wn: ξ4
W'n: 1-ξ4
Wn: ξ2
W'n: 1-ξ2
Wn: ξ3
W'n: 1-ξ3
ρ'
1-ρ'
ρ"
1-ρ"
< 1001
< 1001
< 0001
< 1001
< 1001
< 1001
< 0001
< 1001 >
< 1001 >
< 1001 >
< 1000
< 1001
< 1101
< 1001
< 1001
< 1001
< 1010
< 1001
< 1001 >
< 1001 >
< 1001 >
< 1001 >
< 1000 >
< 1001 >

Fig. 1

it is 0) but the other 15 states form an irreducible, aperiodic chain. Thus, the limiting quantities u_{ijkl} (i.e. the asymptotic probability of being in state $<ijkl>$) exist and are independent of initial conditions. Further, when $\rho' = \rho'' = \rho$ it can easily be shown that u_{ijkl} is also independent of ρ; that is, u_{ijkl} depends only upon the values of ξ_i set by the experimenter.

The data obtained in the experiment (see Suppes and Atkinson [1960, Ch. 11] for detailed information) indicated that the observed probability of an $A_{i,n}$ response approached a fairly stable value over the last 100 trials of the experiment [1]). Also the observed probabilities of an $A_{i,n}$ response given $D_n^{(ij)}$ approached stable values over the last 100 trials of the experiment. The corresponding theoretical predictions for choice behavior can be readily obtained. For example,

$$\lim_{n\to\infty} P(A_{1,n} \mid D_n^{(12)}) = u_{1011} + u_{1010} + u_{1001} + u_{1000}$$
$$+ \tfrac{1}{2}[u_{1110} + u_{1101} + u_{1100} + u_{0011} + u_{0010} + u_{0001} + u_{0000}].$$

And

$$P(A_{i,n}) = \sum_j P(A_{i,n} \mid D_n^{(ij)})P(D_n^{(ij)}).$$

As noted above, u_{ijkl} is a function only of ξ_i when $\rho' = \rho''$. Consequently, in this special case, predictions for $\lim_{n\to\infty} P(A_{i,n} \mid D_n^{(ij)}) = P(A_i \mid D^{(ij)})$ and $\lim_{n\to\infty} P(A_{i,n}) = P(A_i)$ are entirely *a priori* and do not make use of any parameters evaluated from the data.

Table 1 presents the observed response proportions over the last block of 180 trials and the predicted asymptotic values for $\rho' = \rho''$. Overall the model gives a satisfactory account of the mean asymptotic response probabilities when predictions are based solely on experimentally determined parameter values. The correspondence between theory and data could be improved of course if ρ' and ρ'' were estimated from the data and used in generating predictions.

The agreement between these observed and predicted asymptotic response

[1]) On early trials of the experiment the observed values of $P(A_{i,n})$ were approximately $\frac{1}{4}$ which would be expected if the subject initially had no preference among the four responses. The rate at which $P(A_{i,n})$ departs from its initial value and approaches an asymptotic level is of course determined by the reinforcement parameters ρ' and ρ''.

probabilities provides sufficient justification for the type of model construction considered in this paper. However, the model provides a much richer analysis of the experiment than the above results indicate. From the model we can predict not only average performance but also sequential properties of the individual subject's response protocol; i.e. the trial to trial increments and decrements in response probabilities.

TABLE I

Predicted Asymptotic Values and Observed Proportions Over the Last Block of 180 Trials

	Predicted	Observed
$P(A_1 \mid D^{(12)})$	.442	.457
$P(A_1 \mid D^{(13)})$	.355	.445
$P(A_1 \mid D^{(14)})$	.228	.270
$P(A_2 \mid D^{(23)})$	.414	.375
$P(A_2 \mid D^{(24)})$	.286	.273
$P(A_3 \mid D^{(34)})$	.372	.368
$P(A_1)$	.171	.195
$P(A_2)$	.210	.199
$P(A_3)$	.267	.258
$P(A_4)$	.352	.348

It should be emphasized that one of the major contributions of mathematical learning theory has been to provide a framework within which the sequential aspects of learning can be scrutinized. Prior to the development of mathematical models relatively little attention was paid to trial by trial phenomena; at the present time, for many experimental problems such phenomena are viewed as the most basic aspects of learning data.

To indicate the type of sequential predictions that can be obtained from the model, consider the probability of an A_1 response on trial $n+1$ given $D^{(12)}$ on both trial $n+1$ and trial n, a win on trial n, and an A_1 on trial n; namely $P(A_{1,\,n+1} \mid D^{(12)}_{n+1}\, W_n A_{1,n} D_n{}^{(12)})$. To obtain this result we proceed as follows:

$$P(A_{1,n+1}D^{(12)}_{n+1}\ W_nA_{1,\ n}D_n{}^{(12)}) =$$

$$\sum_{j,k} P(A_{1,n+1}D^{(12)}_{n+1}\ C_{j,n+1}W_nA_{1,\ n}D_n{}^{(12)}C_{k,\ n})$$

where $C_{i,\ n}(i = 1, \ldots, 16)$ denotes subject-state i on trial n. In terms of our axioms we may rewrite the sum as

$$\sum_{j,k}P(A_{1,\ n+1} \mid D^{(12)}_{n+1}\ C_{j,\ n+1})P(D^{(12)}_{n+1}\ )P(C_{j,\ n+1} \mid W_nA_{1,\ n}D_n{}^{(12)}C_{k,\ n})$$

$$\cdot P(W_n \mid A_{1,\ n})\, P(A_{1,\ n} \mid D_n{}^{(12)}C_{k,\ n})P(D_n{}^{(12)})P(C_{k,\ n})$$

$$= \tfrac{1}{36}\,\xi_1 \sum_{j,k} P(A_{1,\ n+1} \mid D_n{}^{(12)}C_{j,\ n+1})P(C_{j,\ n+1} \mid W_nA_{1,\ n}D_n{}^{(12)}C_{k,\ n})P$$

$$(A_{1,\ n} \mid D_n{}^{(12)}C_{k,\ n})P(C_{k,\ n}).$$

Each of these quantities in the summation can readily be computed in terms of the axioms. For example, as n becomes large

$$P(A_{1,\ n+1}D^{(12)}_{n+1}\ W_nA_{1,\ n}D_n{}^{(12)}) \overset{n}{\rightarrow}$$

$$\tfrac{1}{36}\,\xi_1\, \{u_{1011} + u_{1010} + u_{1001} + u_{1000} + \tfrac{1}{4}\,[u_{1110} + u_{1101} + u_{1100}]$$

$$+ [\tfrac{1}{2}\,\rho' + \tfrac{1}{4}\,(1 - \rho')]\,[u_{0010} + u_{0001} + u_{0000}]\,\}.$$

To obtain the appropriate conditional probability we divide this result by $\frac{1}{36}\,\xi_1 P(A_1 \mid D^{(12)})$.

In terms of these sequential predictions various procedures can be devised for estimating the reinforcement parameters ρ' and ρ''. Once these parameters have been estimated any theoretical quantity of interest can be computed and goodness-of-fit evaluations made. A consideration of these topics is not appropriate in this paper and the interested reader is referred to Suppes and Atkinson [1960].

At this point it would be nice if we could refer to a list of criteria and a decision rule which would evaluate the model and tell us whether this specific development or similar mathematical models are of any genuine value in analyzing the phenomena of interest to psychologists. Of course, such decision procedures do not exist. Only the perspective gained by refinement and extension of these models with empirical verification at critical stages will permit us to make such an evaluation. Certainly within the last decade almost all learning phenomena have been examined with reference to one or more mathematical models and there is no doubt that these analyses have led to a deeper understanding of the empirical findings. In addition, many new lines of experimentation have resulted directly from the work on mathematical models of learning. In spite of these developments some behavioral scientists maintain that psychology has not yet reached a stage where mathematical analysis is appropriate; still others argue that the data of psychology are basically different from those of the natural sciences and defy any type of rigorous systematization. Of course, there is no definitive answer to these critics. Similar objections were raised to mathematical physics as recently as the late 19th century, and only the brilliant success of the approach silenced opposition. A convincing argument is yet to be made for the possibility that mathematical models in psychology will not enjoy similar success.

University of California, Los Angeles

REFERENCES

1. Audley, R. J. A stochastic model for individual choice behavior. *Psychological Review*, 67 (1960), 1–15.
2. Bower, G. H. Choice-point behavior. Chap. 6 in R. R. Bush and W. K. Estes (Eds.), *Studies in Mathematical Learning Theory*, Stanford University Press, Stanford, 1959.
3. Bush, R. R., and Mosteller, F. A mathematical model for simple learning. *Psychological Review*, 58 (1951), 413–423.
4. Estes, W. K. Toward a statistical theory of learning. *Psychological Review*, 57 (1950), 94–107.
5. Estes, W. K., and Burke, C. J. A theory of stimulus variability in learning. *Psychological Review*, 60 (1953), 276–86.
6. Hull, C. L. *Principles of Behavior*. New York: Appleton–Century–Crofts, 1943.
7. Suppes, P. *Behavioristic foundations of utility*. Technical Report No. 23, Contract Nonr 225(17), Institute for Mathematical Studies in the Social Sciences, Applied Mathematics and Statistics Laboratories, Stanford University, 1959 (to be published in *Econometrica*).
9. Suppes, P., and Atkinson, R. C. *Markov Learning Models for Multiperson Interactions*, Stanford University Press, Stanford, 1960.

SEMANTICS OF PHYSICAL THEORIES

(1) In order to characterize the kind of problem which I wish to consider, I first briefly discuss a situation which is known from classical mechanics. The state of a certain material system at an epoch t is characterized by the values of the *canonical variables* q and p, the number of which may be reduced, for the sake of brevity and simplicity, to one of each type.
For the canonical variables we have the *Hamilton equations:*

$$\frac{dq}{dt} = \frac{\partial H}{\partial p} \qquad \frac{dp}{dt} = -\frac{\partial H}{\partial q}$$

the solution of which may be given by:

$$q = Q(t, a, b), \qquad p = P(t, a, b).$$

I am not interested in the manner in which the Hamilton equations are obtained and solved.
(2) Any physical magnitude m concerning the above material system is supposed to be definable in terms of p, q, and t, and thus we have:

$$m = M(t, a, b).$$

The formula:

$$m[m', t']$$

is taken to express the fact (or the supposition) that a measurement of the magnitude m at the time t' has yielded m' as a value of m.
Stipulation. We shall say that the ordered couple of real numbers $< a', b' >$ *fulfils* the formula $m[m', t']$, if and only if we have $m' = M(t', a', b')$.
Stipulation. We shall say that the conclusion $m[m', t']$ *logically follows* from the premisses $m_1[m_1', t_1'], \ldots, m_k[m_k', t_k']$, if and only if every ordered couple $< a', b' >$ which fulfils all the premisses also fulfils the conclusion.

(3) From the point of view of (non-relativistic) quantum theory a material system is characterized by a *Schrödinger equation:*

$$\frac{d\mathbf{x}}{dt} = \mathbf{Hx},$$

the general solution of which may be given by:

$$\mathbf{x} = \mathbf{X}(t, \mathbf{A}).$$

Physical magnitudes m (including the q's and p's, but not t – hence the non-relativistic character of the theory) are now characterized by linear operators $\mathbf{M}$ on $\mathbf{x}$.

Stipulation. We shall say that the vector $\mathbf{A}'$ *fulfils* the formula $m[m', t']$, if and only if:

$$\mathbf{MX}(t', \mathbf{A}') = m'\mathbf{X}(t', \mathbf{A}').$$

Stipulation. We shall say that the conclusion $m[m', t']$ *logically follows* from the premisses $m_1[m_1', t_1'], \ldots, m_k[m_k', t_k']$, if and only if every vector $\mathbf{A}'$ which fulfils all the premisses also fulfils the conclusion.

(4) In order to show the relevance of the above notions, I briefly discuss the so-called *Einstein-Podolsky-Rosen paradox.* We consider a system formed by two elementary particles. It is supposed that the magnitudes $q_1 - q_2$ and $p_1 + p_2$ are known. Now let us measure q_1 and p_2. Then precise values of q_2 and p_1 can be inferred, notwithstanding the Heisenberg indeterminacy conditions.

In accordance with the above stipulations it can indeed be said that, from the premisses:

$$[q_1 - q_2](q', t'),$$
$$[p_1 + p_2](p', t'),$$
$$q_1(q'', t'),$$
$$p_2(p'', t'),$$

the conclusions:

$$q_2(q'' - q', t')$$
$$p_1(p'' - p', t')$$

logically follow. This is, however, *trivially* so, because of the fact that no vector $\mathbf{A}'$ can be found which fulfils all four premisses.

It is characteristic of quantum theory that it makes allowance for the fact that if we go on collecting information concerning a given physical system, this information tends to become *inconsistent;* this is often expressed by saying that quantum theory makes allowance for *causal anomalies.* In point of fact, if the information at our disposal is inconsistent, then it will be clearly impossible to give a coherent 'historical' account of the way in which the system has been behaving during a certain time-interval.

(5) It need hardly be said that the above remarks do not pretend to give an exhaustive treatment of the intricate problems concerning the logical structure of physical theories, classical or non-classical. They are only meant to suggest the general direction which a more thorough investigation ought to take.

The above semantic stipulations make allowance only for atomic formulas $m[m', t']$. Von Neumann [1]) has shown how sentential connectives are to be taken into account. As in the case of an atomic formula, the vectors $\mathbf{A}'$ which fulfil a given sentential compound will form a linear subspace of the Hilbert space of all vectors $\mathbf{A}'$.

(6) I may add that in my opinion the so-called '*problem of developing quantification theory within the logic of complementarity*' is less urgent than some logicians seem to believe. Curiously enough, as far as I know this problem has been raised exclusively by reviewers in the *Journal of Symbolic Logic* [2]). I wish to emphasize that, for the purpose which in classical logic is served by quantification theory, quantum theory applies the procedures known as composition and second quantization [3]). If second quantization is applied, then the number n of the elementary particles of which a given system is composed is considered as a physical magnitude and thus characterized by a linear operator $\mathbf{N}$. [This is, presumably, connected with the fact that, from the point of view of classical physics, the energetic effect of adding or removing a single object

[1]) J. von Neumann, *Mathematische Grundlagen der Quantenmechanik*, Berlin 1932, pp. 130–134; *cf.* P. Jordan, Quantenlogik und das kommutative Gesetz, in: L. Henkin, P. Suppes, A. Tarski (ed.), *The Axiomatic Method*, Amsterdam (Studies in Logic) 1959, pp. 365–375.

[2]) A. Church on G. Birkhoff and J. von Neumann II 44; C. G. Hempel on H. Reichenbach X 97, and on C. F. von Weizsäcker XXIII65; A. A. Bennett on E. W. Beth XIII 212; J. C. C. McKinsey and P. Suppes on P. Destouches–Février XIX 52.

[3]) H. Weyl, *Gruppentheorie und Quantenmechanik*, 2nd edition, Leipzig 1931, pp. 220–222; J. von Neumann, op. cit., pp. 144, 150; W. Heisenberg, *Die physikalischen Prinzipien der Quantentheorie*, Leipzig 1930, pp. 101 ff.

can be made arbitrarily small and hence, in principle, neglected; thus the *number* of the objects in a system is not properly a physical magnitude. From the point of view of quantum theory this is, of course, not correct.] It is somewhat depressing that, with respect to the problems under discussion, so little progress has been made; it is still more depressing to realize that this may be partly due to the fact that some logicians have raised objections by which prospective authors on the subject have been led astray.

Municipal University, Amsterdam

JEAN-LOUIS DESTOUCHES

SUR LA NOTION DE MODÈLE EN MICROPHYSIQUE

I. LES PREMIERS MODÈLES EN MICROPHYSIQUE

La notion de modèle a pris de l'importance en Physique théorique lors de l'édification de la théorie atomique et de l'ancienne théorie des quanta, lorsqu'il s'est agi de représenter la constitution de l'atome. On a ainsi parlé du '*modèle planétaire de l'atome de Rutherford*'; la force électrostatique de Coulomb ayant même forme analytique que la force de gravitation de Newton, cette représentation était justifiée du point de vue mécanique. Mais un tel modèle se trouvait en contradiction avec les lois de l'électromagnétisme. On a alors parlé du *modèle d'atome de Bohr* où l'image planétaire était conservée mais où les lois de la Mécanique et les lois de l'Electromagnétisme étaient modifiées de façon à rendre cohérent le modèle planétaire et le mettre en accord avec les lois expérimentales sur les spectres.

La structure fine des spectres et l'introduction du spin ont nécessité l'invention de modèles d'atomes plus compliqués dits *modèles vectoriels.* Ces modèles ne rendaient pas compte d'une façon parfaitement satisfaisante des résultats expérimentaux; une fois la Mécanique ondulatoire complètement édifiée, ils ont été tout à fait abandonnés au profit des descriptions de cette Mécanique, qui sont de nature purement prévisionnelle et probabiliste. Les structures qui apparaissaient dans les modèles vectoriels se sont alors trouvées remplacées par des structures algébriques dans un anneau d'opérateurs.

Ces dernières années ont vu un retour de l'utilisation de modèles, mais de modèles voulant respecter les lois établies de la Mécanique ondulatoire, et qui se présentent comme des *modèles hydrodynamiques.*

Dès les débuts de la Mécanique ondulatoire, Madelung avait introduit un modèle hydrodynamique pour cette Mécanique, sous la forme du fluide de probabilité. Ce sont des modèles analogues qui ont été proposés par divers auteurs pour des cas tenant compte du spin ou tenant compte de la relativité. Par exemple, M. Takabayasi [1]) a défini un fluide relativiste associé au fluide de Dirac.

II. LA NOTION DE MODÈLE EN MÉCANIQUE ONDULATOIRE

En effet, le sens du mot '*modèle*' en Mécanique ondulatoire n'est pas tout à fait le même qu'en ancienne théorie des quanta. Dans les deux cas, il s'agit d'images de Mécanique classique qui sont utilisées pour représenter les phénomènes microphysiques; mais en ancienne théorie des quanta, le modèle est constitué par un nombre fini de points matériels et de grandeurs vectorielles, avec des conditions de quantification; en Mécanique ondulatoire ou en théorie de la double solution de Louis de Broglie, il s'agit de modèles hydrodynamiques; dans ce cas, l'élément caractéristique de la théorie est une *fonction*, et à cette fonction on associe, par certaines formules, des fonctions caractérisant un fluide en Mécanique classique. Par exemple, pour un corpuscule sans spin, on dispose en Mécanique ondulatoire, d'une fonction d'onde complexe ψ et on lui associe un fluide classique, appelé fluide de probabilité, caractérisé par deux fonctions réelles, une densité ϱ et un potentiel des vitesses Φ, fonctions définies par

$$\varrho = |\,\psi\,|^2, \quad \Phi = -\,\frac{1}{m}\arg\psi,$$

où 'arg' signifie 'argument imaginaire de'.

Comme conséquence de l'équation d'ondes pour la fonction ψ, la fonction ϱ satisfait à une équation de continuité hydrodynamique et la fonction Φ à l'équation du potentiel des vitesses qui s'obtient en hydrodynamique à partir de l'équation d'Euler, en remontant des gradients aux potentiels.

Dans la théorie de la double solution, la fonction ψ continue à représenter les prévisions comme en Mécanique ondulatoire usuelle, et en outre une fonction u obéissant aussi à une équation d'ondes représente d'une façon objective le corpuscule. Par les mêmes formules que ci-dessus, en remplaçant ψ par u, on peut associer à un corpuscule représenté par une onde u, un fluide classique; c'est ce que l'on peut appeler *le modèle hydrodynamique du corpuscule*. De nombreux développements ont été effectués à partir de cette image hydrodynamique d'un corpuscule.

On peut se demander quelle est la valeur de tels modèles. Certains auteurs vont jusqu'à croire que ces modèles décrivent réellement les

[1]) Takabayasi, *Progress theor. Phys.*, t. 8, 1952, p 143–182; t. 11, 1954, p. 341–373; t. 14, 1955, p. 283–302; *Nuovo Cimento* 10e S; t. 3, 1956, p. 233–241; *Suppl. Progress theor. Phys.* no. 4, 1957.

processus physiques qu'ils représentent. D'autres pensent qu'ils fournissent seulement des images concrètes permettant de fixer les idées, pour servir de base à des constructions théoriques.

On doit constater que les modèles de l'ancienne théorie des quanta ont fourni une aide efficace pour le développement de la théorie, mais qu'ils ont complètement disparu en raison même du dépassement de cette théorie par la formation de la Mécanique ondulatoire. Dans les circonstances actuelles, les modèles hydrodynamiques ont été une aide initiale; par exemple, ils m'ont fourni un moyen aisé de vérifier la cohérence des hypothèses mises à la base de la théorie de la double solution [1]) en me permettant de prouver par la construction d'un modèle hydrodynamique, la compatibilité de ces hypothèses. Ici, le modèle se trouve utilisé en Physique théorique un peu comme en Logique, et non plus simplement comme support représentatif d'idées intuitives. Mais d'autre part, on peut se demander si l'image hydrodynamique d'un corpuscule n'est pas de nature purement formelle, due au fait qu'à partir d'une fonction complexe u obéissant à une équation d'ondes, on peut fabriquer des fonctions qui obéissent à des équations de l'hydrodynamique classique. C'est d'ailleurs justement cette dérivation mathématique qui permet d'utiliser le modèle hydrodynamique pour des démonstrations de cohérence.

III. LA NOTION DE MODÈLE EN THÉORIE FONCTIONNELLE

En critiquant la notion de système physique et la distinction d'un système au sein de l'Univers, on est conduit à représenter un corpuscule par une certaine fonction de point de l'espace-temps $u(P, t)$, et ceci est le point de départ de la théorie fonctionnelle des corpuscules. A partir de cette fonction u, on peut dériver un modèle hydrodynamique [2]), mais ceci

[1]) J. L. Destouches, *L'onde u et le fluide associé dans la théorie de la double solution de M. Louis de Broglie (J. Phys. Rad.*, t. 16, 1955, p. 81–85); *Sur la compatibilité de certaines hypothèses de la théorie de la double solution de M. Louis de Broglie* (*J. Phys. Rad.*, t. 16, 1955, p. 86–91); *La quantification en théorie fonctionnelle des corpuscules* (Gauthier-Villars, Paris, 1956).

[2]) J. L. Destouches, *La quantification en théorie fonctionnelle des corpuscules* (Gauthier-Villars, Paris 1956); *Quantization in the functional theory of particles* (Nuovo Cimento, suppl. vol. III série X, 1956); *Corpuscules et Champs en théorie fonctionnelle* (Gauthier-Villars, Paris, 1958).
J. L. Destouches & F. Aeschlimann, *Les systèmes de corpuscules en théorie fonctionnelle* (Hermann, Paris, 1959).

paraît purement formel et sert seulement à fournir une image intuitive d'un corpuscule comme une goutte de fluide ou encore comme un nuage de fluide, ces images hydrodynamiques ne semblant par correspondre à la réalité profonde. Au contraire, la représentation fonctionnelle paraît douée d'une signification physique essentielle parce qu'elle est introduite par des considérations directement physiques et non pas par une analogie formelle. Et dans le développement de la théorie fonctionnelle, les images hydrodynamiques ont perdu leur importance, de même qu'en Mécanique ondulatoire usuelle, le modèle hydrodynamique de Madelung.

On pourrait chercher à définir avec précision ce que l'on doit appeler un '*modèle*' en Physique, de même qu'en Sémantique on définit exactement ce qu'est un modèle. Il semble qu'on ne pourra y parvenir qu'en faisant appel à des comparaisons de structures. Mais à première vue de telles comparaisons paraissent difficiles à réaliser parce qu'elles entraînent à présupposer l'existence d'une certaine structure décrivant exactement la réalité physique, et que l'on dispose d'une autre structure constituant le modèle, qui aura avec la structure réelle une relation d'homomorphisme. Selon la position philosophique que l'on adopte, ou bien on ne suppose pas de structure réelle, et alors la définition du modèle ne s'applique pas, ou bien on adopte une position rationaliste sommaire et on admet l'existence d'une structure réelle, mais comme on ne la connaît pas effectivement, la définition du modèle est métaphysique.

On pourrait alors élargir cette définition trop sommaire du modèle en remarquant qu'en physique théorique, nous décrivons les phénomènes au moyen d'une théorie qui, une fois construite, peut être considérée comme un système formel ayant une structure mathématique qui constitue la structure formelle de la théorie, et faire appel ensuite à des structures successives approximant la description de la réalité. Quand on passe d'une théorie à une théorie meilleure, la structure de la théorie est changée tout en conservant une partie de celle de la théorie primitive. D'autre part, un modèle possède aussi une structure formelle. Pour que soit justifié son nom de modèle, il faut que sa structure présente un homomorphisme avec la structure théorique de première approximation décrivant la réalité physique. Si l'homomorphisme se maintient avec une structure de deuxième approximation, on aura un *modèle adéquat*. Un *modèle efficace* sera un modèle qui suggère la construction d'une structure de deuxième approximation. Le plus souvent, avec cette deuxième approximation, l'homomorphisme n'existera plus et le modèle ne sera plus utilisable, comme ce fut le cas pour les modèles de l'ancienne théorie des quanta.

A propos de la notion de modèle, nous nous trouvons devant la même difficulté que pour toutes les notions fondamentales de la Physique théorique. Celles-ci ne sont jamais qu'à peu près adéquates, approchant la réalité physique de beaucoup moins près qu'on ne le croyait autrefois, et la difficulté est toujours de parler d'une façon précise et rigoureuse sur des notions qui ne peuvent être qu'approximatives.

C'est toujours le même difficile problème qui apparaît avec les 'mathématiques de l'à peu près', ou encore avec la question du continu physique. On peut dire que le continu mathématique est un modèle du continu physique, qui peut se trouver adéquat dans un grand nombre de cas, mais qui dans un certain nombre d'autres, ne fournira pas une description physique adéquate. Par exemple, un corpuscule en tant qu'insécable, peut être décrit par un point géométrique, mais cette description ponctuelle ne se montre pas adéquate pour décrire tous les aspects physiques du corpuscule, et la description fonctionnelle d'un corpuscule au moyen d'une fonction u, peut être considérée comme liée au fait que le modèle constitué par le continu mathématique est imparfaitement adéquat à décrire le continu physique. La représentation fonctionnelle contient justement l'interpénétration que réclame la continuité physique, tout en distinguant les éléments que sont les corpuscules.

IV. CONCLUSION

Pour conclure, nous dirons qu'un modèle en Microphysique est constitué par un système mécanique obéissant aux lois de la Mécanique classique, soit de la Mécanique ponctuelle, soit de la Mécanique des milieux continus. Un modèle ne constitue pas une théorie mais fournit une image concrète du système microphysique que l'on étudie dans une théorie, soit déjà faite, soit à édifier. Une théorie nouvelle ne pourra être édifiée qu'après plusieurs tentatives successives. Chacune de ces étapes possèdera une structure mathématique, et la structure du modèle devra présenter un homomorphisme au moins avec la structure de la première tentative théorique. Un modèle peut avoir une utilité en suggérant des constructions théoriques qui constitueront les étappes ultérieures de l'édification de la théorie cherchée; mais le plus souvent l'homomorphisme du modèle ne subsistera pas dans ses étapes ultérieures et alors le modèle cessera de jouer un rôle heuristique.

Institut Henri Poincaré, Paris

ROBERT FEYS

MODÈLES À VARIABLES DE DIFFÉRENTES SORTES POUR LES LOGIQUES MODALES M″ OU S 5

0.1 La logique modale a été recréée à notre époque, à l'occasion de certains 'paradoxes' de l'implication matérielle, que Lewis entendait éliminer. Mais sa logique modale offre à son tour quelque chose de paradoxal: ses interprétations ont quelque chose d'essentiellement instable; elles paraissent se dérober lorsque'on tente de les fixer catégoriquement. En logique modale l'affirmation simple 'il pleut' se situe dans un présent ambigu; l'affirmation 'il pleuvait hier' semble affirmer la vérité présente d'une réalité passée, et on ne sait si 'l'étoile du matin' et 'l'étoile du soir' doivent être effectivement distinguées. Ces difficultés d'interprétation semblent annoncer ce que nous appelerions des antinomies structurelles, l'impossibilité de formuler des règles cohérentes de structure.

0.2 Or la construction d'un modèle fournirait une preuve de cohérence. A défaut de posséder une interprétation satisfaisante en mots, il peut y avoir intérêt à se guider sur un équivalent formalisé de la logique modale ou de certaines logiques modales. Cet équivalent sera libre des ambiguïtés du langage et il constituera un modèle, en tant qu'il représente adéquatement (pour les besoins de la déduction) la structure des entités logiques considérées.

0.3 L'interprétation à laquelle nous recourons est bien connue. Elle envisage des affirmations qui sont valables dans certains cas, dans certaines circonstances: l'affirmation est néccessaire si elle est valable en tout cas, elle est possible si elle est valable dans certains cas.

Plus concrètement considérons un évènement réalisable à divers moments: l'évènement sera nécessaire q'il a lieu à tout moment, et possible s'il a lieu à certains moments. Nous pourrons également considérer divers 'mondes', ce qui nous introduit pas forcément dans une fantasmagorie de mondes imaginaires. Soit une fédération d'états, chaque état sera un monde qui a sa législation à lui. Alors une loi sera possible si elle est en vigueur dans certains états, nécessaire si elle est en vigueur dans tous les états.

0.4 La logique que nous adopterons comme modèle est donc une structure particulièrement simple: les propositions modales y sont isomorphes avec des propositions générales (universelles et particulières) de la logique classique.

I. LOGIQUES DES PROPOSITIONS TF_1^1 ET $M''F_0$

A. LOGIQUE F_1^1

Nous considérerons une logique TF_1^1, qui comporte des variables T (1.01 et 1.03) qui a la même structure qu'une logique AF_1^1 de ler ordre et ler degré et qui est équivalente à la logique des propositions $M''F_0$ (F_0 désigne la logique des propositions, M″ signifie qu'il s'agit de la logique modale M″ ou S 5).

1.0 LOGIQUE TF_1^1: RÈGLES DE STRUCTURE

1.01 Les variables p, q, n, p′, q′... sont des variables d'évènements ou d'éventualités. Elles sont désignées par P, Q, N... Ultérieurement des majuscules pourront désigner également des abstraits.
La variable t est une variable de cas, une variable casuelle, elle forme une catégorie désignée par t.
Nous utiliserons fréquemment une expression pour se désigner elle-même, lorsqu'elle figure dans une phrase du langage parlé usuel. Nous venons p.ex. de parler de variables t, p, q, et pas des variable 't', 'p' 'q'.
1.02 Une expression PT, où P est une variable, est une proposition atomique. Elle pourra se lire: P est vrai dans le cas T.
1.03 Nous assignerons aux expressions de TF_1^1 des types conformément à la méthode de Church (STT). Nous établirons de la sorte que les règles de structure de TF_1^1 ne sont pas inconsistantes, au moins sous ce rapport.
La variable t n'est pas du type ι de Church; ce n'est pas une variable d'individu. Nous aurons affaire à des expressions 'mehrsortig'. A la variable t sera signé un type τ.
Une expression PT est d'un type que nous appelons υ et qui est le type des propositions des logiques T. Par application à une variable T, une variable P donne une expression PT du type υ. La variable P sera du type $\upsilon\tau$.

1.04 Une variable P ne peut être affirmée; elle n'énonce pas une proposition. Seule une affirmation PT relative à un cas donné peut, en logique T, être affirmée comme proposition atomique.
1.05 L'expression At signifie 'pout tout cas t'; l'expression At pt signifie 'pour tout cas t, on a pt' et constitue une proposition (générale).
L'expression Et signifie 'pour certains cas t', l'expression Et pt signifie 'pour certains cas t, on a pt' et constitue une proposition (générale).
1.06 Nous n'introduitons pas dans le langage TF de constante désignant tel cas déterminé.
1.07 Il n'y a qu'une variable T. Nous admettrons qu'une expression comme AtAt pt est bien formée et que le 't' de 'pt' est lié par le le quantificateur à gauche de 'pt'.
1.08 En conclusion toutes les expression de TF_1^1 sont relatives à des cas t, mais ceux-ci sont laissés dans l'indétermination. Une proposition M contient une variable t libre ou liée.

1.1 RÈGLES DE DÉDUCTION ET DÉFINITIONS DE TF_1^1

1.10 Tous les postulats et toutes les définitions de la logique des propositions AF_0 (logique non modale) sont valables pour les propositions TF_1^1.

1.11 Et pt pour $\sim$ At $\sim$ pt
1.12 pt $\underset{t}{\rightarrow}$ qt pour At (pt $\rightarrow$ qt)
1.13 pt $\underset{t}{\leftrightarrow}$ qt pour At (pt $\leftrightarrow$ qt)
1.14 $\vdash$ At pt $\rightarrow$ pt
1.15 $\vdash$ At (pt $\rightarrow$ qt) $\rightarrow$ (At pt $\rightarrow$ At qt)
1.16 $\vdash$ Et At pt $\rightarrow$ At pt
1.17 Si $\vdash$ PT alors AT PT

1.2 THÉORÈMES DE TF_1^1

1.2 Partant des postulats ci-dessus on déduira des théorèmes correspondants à ceux de la logique AF_1^1 de premier ordre et premier degré.

On a par ex. $\vdash$ At qt $\rightarrow$ (pt $\underset{t}{\rightarrow}$ qt)

En effet: (1) $\vdash$ qt $\rightarrow$ (pt $\rightarrow$ qt) 1.10
(2) $\vdash$ At [qt $\rightarrow$ (pt $\rightarrow$ qt)] 1.18
(3) $\vdash$ At qt $\rightarrow$ At (pt $\rightarrow$ qt) 1.15
(4) $\vdash$ At qt $\rightarrow$ (pt $\underset{t}{\rightarrow}$ qt) 1.12

1.22 Remarquons que par un usage répété de 1.28 nous pouvons fréquemment écrire $\underset{t}{\rightarrow}$ ou $\underset{t}{\leftrightarrow}$ où la logique usuelle écrit $\rightarrow$ ou $\leftrightarrow$. P. ex. nous pouvons ajouter à la démonstration précédente:

(5) $\vdash$ At [At qt $\rightarrow$ (pt $\rightarrow$ qt)] 1.18
(6) $\vdash$ At qt $\underset{t}{\rightarrow}$ (pt $\underset{t}{\rightarrow}$ qt) 1.12

Nous ne discutons pas pour le moment l'intérêt ou le caractère plutôt banal de cette dernière possibilité.

B. ABSTRAITS DANS LE SYSTÈME TF_1

Les abstraits d'une logique des prédicats peuvent être interprétés de façon cohérente somme désignant, soit des propriétés prises abstraitement, soit des ensembles ordinairement représentables comme collections d'individus. La traduction d'un énoncé en termes d'abstraits permettra de formuler explicitement sous l'un ou l'autre de ces aspects la signification de cet énoncé, nous disons la 'signification', la 'Bedeutung', car si les abstraits s'interprètent comme ensembles, comme 'extensions' ils conduisent de ce fait à quelque chose d'analogue à une interprétation extensionnelle.

1.3 ABSTRAITS

1.31 Arrêtons-nous d'abord aux abstraits **t** M. Soit M une proposition de TF_1^1, l'abstrait **t** M exprimera l'aspect commun à toutes les affirmations M ou l'ensemble des cas t tels que M.

1.32 On a l'équivalence

$$\vdash (\mathbf{t}\ M)\ t \leftrightarrow M$$

où t est une indéterminée.

1.33 Un abstrait **t** M constitue une valeur d'une variable p.

1.34 Les abstraits pourraient être exprimés à l'aide des notations de la

lambda - conversion; alors λ t M pourrait être une expression bien formée même si M n'était pas une proposition de TF_1^1.
1.35 Puisque un abstrait **T** M désigne un enxemble, les fonctions d'abstraits pourront être représentées par des ensembles qui sont fonctions d'autres ensembles, p. ex. par des ensembles de points, cercles d'Euler, diagrammes de Venn ou, plus simplement par des rapports entre portions de ligne, à la manière de Leibniz.
1.36 Introduisons ici une notation semblable à celle de Church (STT). En général si M est du type α et si un terme ξ est du type β, $\lambda\xi$ M est du type $\alpha\beta$. Un abstrait ξ M, où M est une proposition, sera du type $\upsilon\beta$. En particulier un abstrait **t** M sera du type $\upsilon\tau$.
L'application d'un abstrait de type $\alpha\beta$ à un terme β sera du type α. Ceci peut être rappelé en désignant le type en question comme $\alpha\beta{:}\beta$.
De même si $\delta{:}\beta{:}\gamma$ est α, δ est $\alpha\beta\gamma$.
1.37 Signalons une difficulté à laquelle se heurte la théorie des abstraits. Il semble inévitable de penser que si M est une proposition quelconque de TF_1^1, **t** M sera une expression bien formée qui constitue un abstrait. D'ordinaire M contiendra t comme (seule) variable libre, ce n'est que dans une extension de TF (voir 1.80) que M pourrait contenir une variable de cas autre que t. Mais que dire d'un abstrait où M est une proposition à quantificateur At et Et? L'abstrait sera p. ex. de la forme **t** At pt ou **t** Et pt. On peut formuler une règle de *réduction* telle que (**t** At pt) t, soit At pt. Mais on tomberait alors dans *l'extension* de TF_1^1.
1.39 Bornons-nous à signaler la possibilité de poser des abstraits (abstraits de 2e ordre) **P** M et en particulier **p** pt. Ces derniers exprimeront ce qu'il y a de commun aux propositions concernant un même cas t: ou encore des événements réalisés dans un cas t.

1.4 FONCTIONS D'ABSTRAITS T M

Nous pouvons comme en logique des classes considérer des fonctions d'abstraits **T** M.
1.40 p est sustituable à **t** pt.
Nous pouvons définir
1.41 — p pour **t** pt
1.42 p $\cap$ q pour **t** (pt qt)
1.43 p $\cup$ q pour **t** (pt qt)

Nous pourrions définir une classe totale et une classes nulle de cas, comme suit:

1.44 $1_{\upsilon\tau}$ pour **t** (At pt)

1.45 $0_{\upsilon\tau}$ pour **t** (At ~ pt)

Nous définirons de même:

1.46 $p \subseteq q$ pour pt $\underset{t}{\rightarrow}$ qt

1.47 $p = q$ pour pt $\underset{t}{\leftrightarrow}$ qt

c'est ici qu'apparaît la portée de l'implication nécessaire et de l'équivalence nécessaire. Une équivalence (implication) nécessaire se traduit par une identité d'abstraits, une identité (inclusion) de notions abstraites.

1.5 THÉORÈMES EN TERMES DE CLASSES DE CAS

Nous pouvons donc considérer une implication ou une équivalence nécessaire comme exprimant l'inclusion ou l'identité de deux 'événements' ou de deux ensembles de cas.

1.51 Nous appliquerons ce mode d'expression à toutes les lois logiques du No. 1.2.

La loi ⊢ At qt → (pt $\underset{t}{\rightarrow}$ qt)

sera ⊢ Aq → ($p \subseteq q$)

1.52 Avec des superpositions d'implications ou équivalences nécessaires nous serrons conduits à des notations et interprétations plus compliquées. P. ex.

la loi ⊢ At qt $\underset{t}{\rightarrow}$ (pt $\underset{t}{\rightarrow}$ qt)

serait ⊢ **t** (At qt) ⊆ **t** ($p \subseteq q$)

Mais ces complications de notations semblent refléter le caractère plus ou moins surperfétatoire de ces relations nécessaires surajoutées.

C. LA LOGIQUE M″ F_0

1.6 CORRESPONDANCE ENTRE M″ F_0 ET TF_1^1

1.60 La correspondance entre ces deux logiques doit se régler sur l'interprétation intuitive 0.2. Les énonciations des deux logiques doivent être déductivement équivalentes, mais celles de la logique M″F_0 ne mentionneront aucune variable de cas. TF_1^1 doit donc être telle que moyennant une règle d'abréviation sans ambiguïté, elle devienne M″F₀. Les expressions de TF_1^1 contiendront une variable unique, la variable t, et celle-ci ne sera *pas exprimée* dans $M''F_0$. Il ne pourra être question de la biffer purement et simplement; c'est pourquoi les variables de $M''F_0$ ne seront pas des variables p, q, n, mais bien *p*, *q*, *n*. C'est notamment en quoi les expressions en termes de classe des nos 1.3 à 1.5. différeront des expressions de $M''F_0$.

1.61 On remplacera les symboles de la ligne supérieure par ceux de la ligne inférieure, donc:

$$\text{pt, qt, nt} \ldots \text{At Et} \underset{t}{\rightarrow} \underset{t}{\leftrightarrow} \text{ par}$$

$$p \quad q \quad n \ \ldots \Box \ \Diamond \Rightarrow \ \Leftrightarrow$$

1.62 Concluons: toutes les expressions de $M''F_0$ restent relatives à des cas, mais à des cas indéterminés. Et c'est pourquoi les cas peuvent être sous-entendus.

1.7 POSTULATS ET THÉORÈMES DE $M''F_0$

En appliquant la correspondance 1.6 aux postulats du No. 1.1, ceux-ci deviennent:

1.70 Tous les postulats et toutes les définitions de la logique (non modale) des propositions AF_1^1 sont valables pour les propositions de TF_1^1. Et on a:

1.71 $\Diamond p$ pour $\sim \Box \sim$ p

1.72 $p \Rightarrow q$ pour $\Box$ (p $\rightarrow$ q)

1.73 $p \Leftrightarrow q$ pour $\Box$ (p $\leftrightarrow$ q)

1.74 $\vdash \Box p \rightarrow p$

1.75 $\vdash \Box (p \rightarrow q) \leftarrow (\Box p \rightarrow \Box q)$

1.76 $\vdash \Diamond \Box p \rightarrow \Box p$

1.77 Si $\vdash P$ alors $\vdash \Box P$.

A partir de ces postulats on peut démontrer tous les théorèmes d'un système $M''F_0$ déductivement équivalent au système S 5 de Lewis ou au système M″ de von Wright. Dans ce système se déduoront p. ex. des théorèmes équivalents aux théorèmes 1.21 et 1.22 et ce à l'aide d'une démonstration correspondante aux démonstrations de 1.2.

Soit donc à démontrer $\vdash \Box q \Rightarrow$ (p $\Rightarrow$ q).

(1) $\vdash q \rightarrow (p \rightarrow q)$ 1.70

(2) $\vdash \Box\, [q \rightarrow (p \leftarrow q)]$ 1.77
(3) $\vdash \Box\, q \rightarrow \Box\, (p \rightarrow q)$ 1.75
(4) $\vdash \Box\, q \rightarrow (p \Rightarrow q)$ 1.72
(5) $\vdash \Box\, [\Box\, q \rightarrow (p \Rightarrow q)]$ 1.77
(6) $\vdash \Box\, q \Rightarrow (p \Rightarrow q)$ 1.72

D. EXTENSIONS DE TF_1^1 ET $M''F_0$

1.8 EXTENSIONS DE TF_1^1

Un des aspects intéressants de TF_1^1 est qu'il se prête à diverses extensions.
1.80 Le seul fait d'introduire diverses variables de cécas donnerait lieu à toutes les complexités de la logique des relations.
1.81 Supposons que les variables casuelles soient ordonnées; elles pourraient exprimer divers moments dans la suite des temps ce qui permettrait de traduire des phénomènes se succédant dans le temps (phénomènes électriques, décharges nerveuses).
1.82 Il est aisé de spécifier le nombre de 'cas' possibles dans un système; la spécification peut se faire à l'aide de Zahlformeln (supposant une relation d'identité entre cas).
Dès les origines de la logique des propositions on s'est servi de ce qui est, en fait, des Zahlformeln pour définir une logique modalse se réduisant à la logique classique. Si nous recourons à la notation par classes (1.5) la condition de réduction serait:
1.83 $\sim (p = 1_{\upsilon\tau}) \leftrightarrow (p = 0_{\upsilon\tau})$
1.84 Des logiques modales à n cas ($n > 1$) seraient au contraire comparables à des logiques plurivalentes à $n + 1$ valeurs.
1.85 Jusqu'à présent il nous est impossible d'énoncer que l'événement se produit dans tel cas déterminé ou constant. Il nous est impossible p. ex. d'énoncer que l'événement se produit dans 'le cas réel'. Si nous introduisons un cas réel t_r, l'équivalent du p de la logique classique sera pt_r.
1.86 Il n'est pas exclu d'adjoindre des propositions Pt_r aux propositions PT, mais il faudrait poser des axiomes pour des fonctions de vérité de pt_r et de pt.

1.9 EXTENSIONS DE $M''F_1^1$

Dans quelle mesure les extensions de TF_1^1 pourraient-elles être traduites dans le système $M''F_0$, système proprement modal, donc sans variables casuelles?

1.90 Il semble que 1.80 permettrait une règle de réduction plus ou moins subtile et que, avec un usage limité variables T supplémentaires, on éliminerait de la sorte des difficultés des No. 1.37 et 2.36 ci-dessous.

1.91 Les extensions du No. 1.81 paraissent impossibles à traduire de la sorte; pour que la variable libre de cas puisse être omise il paraît essentiel qu'elle soit unique, si elle n'est pas éliminable dans TF^1_1 déjà.

1.92 La réduction de la logique modale à une logique classique peut se traduire, comme dans 1.83, par des axiomes d'équivalence. Elle pourrait aussi s'exprimer (c'est ce qui se fait déjà aux origines de la logique propositionnelle) en remplaçant les notations de la logique modale par celles de la logique classique.

1.95 Il n'est pas exclu d'employer les variables de la logique classique pour exprimer l'affirmation dans le cas réel. Mais si comme dans ce texte, nous réservons les italiques pour les variables de $M''F_0$ il nous faudra d'autres caractères p. ex. des minuscules grasses pour les variables du "cas réel".

1.96 On se demandera peut être pourquoi nous n'avons pas pris comme équivalent du p de $M''F_0$ la proposition Pt_r. Dans cette interprétation qui paraît première vue plausible, l'axiome 1.28 ne serait plus valide. Or cet axiome est valide dans S 5 – ainsi que dans S 4 (système équivalent à M') et dans le 'système t' équivalent à M. La question est ouverte de savoir quelle équivalence adopter pour des systèmes comme S 2 et S 1.

II. LOGIQUES DES PREDICATS TF^1_2 ET $M''F^1_1$

Nous considérons maintenant une logique TF^1_2 de 1er ordre et de 2e degré, que nous verrons être comparable à une logique équivalente à la logique modale des classes (de 1er ordre) $M''F^1_1$. Nous mentionnerons des logiques TF^1_{n+1} ($n > 1$), équivalentes à des logiques $M''F^1_n$.

A. LA LOGIQUE TF^1_2

2.0 LOGIQUE TF^1_2 – RÈGLES DE STRUCTURE

2.01 Les variables de la logique TF^1_e sont:

a) des variables d'éventualités p, q, n . . .
b) la variable de cas t,
c) des variables a, b, c, d, . . . que nous appellerons des variables prédicatives.
d) des variables x, y, z . . . sue nous appellerons des variables individuelles.

Nous emploierons comme désignations des variables les majuscules correspondantes (Ultérieurement une majuscule A, B . . . pourra désigner également un abstrait.)

2.02 Une expression (AX)T est une propositon atomique de TF_2^1, si A est une variable. Les propositions PT sont également des propositions atomiques de TF_2^1.

2.03 (AX)T peut s'écrire, par association à gauche, AXT.

2.04 L'expression (AX)T est du type υ. L'expression AX sera du type $\upsilon\tau$.

Designons par η le type des variables, individuelles – nous évitons d'identifier ces variables avec les variables individuelles de la logique classique. Alors, puisque AX est du type $\upsilon\tau$, A sera d'un type a tel que $a{:}\eta{:}\tau$ soit υ. Une variable A sera donc du type $\upsilon\eta\tau$ ou $(\upsilon\eta)\tau$.

2.05 Nous lisons axt: l'individu (éventuel) x possède le prédicat (éventuel) a dans le cas t.

2.06 On voit que la proposition axt dont l'interprétation est analogue à celle de la proposition *ax* en logique classique, possède une structure analogue à celle d'une proposition relative classique *rxy*.

Nous pouvons écrire axt sous la forme ătx, de même que en logique classique *řyx* équivaut à *rxy*.

Nous pouvons également écrire axt, sous la forme a (x, t) attribuant le prédicat a au couple x, t.

2.07 Les propositions atomiques de TF_2^1 peuvent être affectées de quantificateurs AX (pour tout X) ou EX (pour certains X), X étant une variable individuelle.

Soit Q le signe A ou le signe E.

Nous devons considérer deux sortes de propositions générales dans TF_2^1. D'une part des propositions Qx M (quantificateurs simples), d'autre part des propositions Qx Qt M ou Qt Qx M (quantificateurs doubles). Telles seront les propositions Ax At (axt $\rightarrow$ bxt) et Ax Et (axt $\leftrightarrow$ bxt)

2.1. POSTULATS DES LOGIQUES TF^1_{n+1} ($n \geqslant 1$)

Ces postulats comporteront:

2.11 Les postulats de la logique AF^1_1. Ce sont les postulats usuels des propositions avec quantificateurs pour autant qu'ils ne concernent qu'une variable liée.

2.12 Les postulats 1.1 de la logique TF^1_1.

2.13 ⊢ At Ax M ↔ Ax At M.

Les logiques TF^1_{n+1} où $n > 1$ comprendraient en outre le postulat.

2.14 ⊢ Ax Ay M ↔ Ay Ax M.

2.2 THÉORÈMES DE LA LOGIQUE TF^1_{n+1} ($n \geqslant$)

2.21 Les théorèmes où ne figurent que les quantificateurs simples (voir 2.07) sont exactement les mêmes que ceux de AF^1_1, sauf que les propositions atomiques sont de la forme AXT.

2.22 Il en serait de même des théorèmes de logiques TF^1_{n+1} ($n \geqslant 1$) du moment qu'ils ne comportent pas de variables T. Ces théorèmes seraient exactement ceux de AF^1_n sauf que les propositions atomiques pourraient comporter n variables individuelles.

2.23 Dans la logique TF^1_2 les théorème à quantificateurs doubles Ax At ou At Ax, ou bien Ex Et ou Et Ex seraient exactement semblables à ceux de 2.21.

On sait en effet (2.13) que ⊢ Ax At axt ↔ At Ax axt et que par suite ⊢ Ex Et axt ↔ Et Ex axt.

2.24 De même pour une logique TF^1_{n+1} ($n > 1$) par rapport aux théorèmes du No. 2.22.

2.25 Les seules légères différences avec la logique AF^1_1 concerneraient les théorèmes où figurent des quantificateurs Ax Et, Et Ax, Ex At, At Ex. Ici en effet on doit tenir compte de ce qu'on a ⊢ Ex At axt → At Ex axt et ⊢ Et Ax axt → Ax Et axt et que l'on n'a pas les implications inverses.

2.26 ⊢ Ex (bxt → pt) → (Ax bxt → pt)

x n'étant pas libre dans p.

En effet ce théorème équivaut à

⊢ Ex At (bxt → pt) → At Ex (bxt → pt)

mais l'inverse équivaudrait à

At Ex (bxt → pt) → Ex At (bxt → pt)

2.27 $\vdash \mathrm{Ex}\,(pt \to bxt) \to (pt \to \mathrm{Ex}\,bxt)$
x n'étant pas libre dans p.
En effet ce théorème équivaut à

$$\vdash \mathrm{Ex\,At}\,(pt \to bxt) \to \mathrm{At\,Ex}\,(pt \to bxt)$$

Mais l'inverse équivaudrait à

$$\mathrm{At\,Ex}\,(pt \to bxt) \to \mathrm{Ex\,At}\,(pt \to bxt).$$

B. ABSTRAITS DANS LA LOGIQUE TF_2^1

2.3 DIVERSES ESPÈCES D'ABSTRAITS

2.31 Outre les abstraits **tM** déjà mentionnés nous avons à traiter de deux formes d'abstraits:
a) les abstraits simples **x** M
b) les abstraits doubles **xt** M
2.32 Un abstrait **xt** M constitue une valeur d'une variable a.
Un abstrait **x** M n'est pas valeur d'une des variables qui ont été introduites jusqu'ici.
2.33 Un abstrait **xt** M étant comparable à une relation binaire pourra être représenté par un tableau à double entrée, de la façon usuelle.
2.34 Un abstrait **x** M est du type $\upsilon\eta$
Un abstrait **xt** M est du type $\upsilon\eta\tau$.
2.35 La proposition M peut contenir x comme variable libre, y comme variable libre ou dans des quantificateurs.
Commes les variables individuelles sont soumises aux même postulats qu'en logique classique, l'expression **x** Ax axt n'est pas bien formée. L'expression ŷ axt, l'expression ŷ Ax Axt sont bien formées.
2.36 Les abstraits **x** At Axt, **x** Et axt sont bien formés. Il se traduisent par 'posséder nécessairement la propriété a' 'posséder possiblement la propriété a'. Mais **xt** At axt, **xt** Et axt ne sont pas bien formés (voir 1.37).

2.4 FONCTIONS D'ABSTRAITS X M

2.41 — (**x** M) pour **x** $\sim$ M
2.42 (**x** M) $\cap$ (**x** M) pour **x** (M $\wedge$ M′)
2.43 (**x** M) $\cup$ (**x** M) pour **x** (M $\vee$ M′)

2.44 $1_{\upsilon\eta}$ pour **x** (Ax ∼ ax)
2.45 $0_{\upsilon\eta}$ pour **x** (Ax ∼ ax)
2.46 (**x** M) ⊆ (**x** M′) pour M $\underset{x}{\rightarrow}$ M′
2.47 (**x** M) = (**x** M′) pour M $\underset{x}{\leftrightarrow}$ M′.

Une implication ou équivalence 'pour tout x' exprime un rapport entre propriétés considérées relativement à des cas donnés.

2.5 FONCTIONS D'ABSTRAITS **x t** M

2.50 a est substituable à **x t** axt.
Nous définissons:
2.51 — a pour **xt** ∼ axt
2.52 a ∩ b pour **xt** (axt bxt)
2.53 a ∪ b pour **xt** (axt bxt)
2.54 $1_{\upsilon\eta\tau}$ pour **xt** (Ax At axt)
2.55 $0_{\upsilon\eta\tau}$ pour **xt** (Ax At axt)
2.56 a ⊆ b pour axt $\underset{xt}{\rightarrow}$ bxt
2.56 a = b pour axt $\underset{xt}{\leftrightarrow}$ bxt.

Une implication ou équivalence doublement générale exprime un rapport entre propriétés abstraites, sans référence à un cas donné ou à tous les cas ou à certains cas.

C. LA LOGIQUE $M''F_1^1$

2.6 TRADUCTION

Il faudra donc traduire les expressions de la logique TF_2^1 en expressions d'une logique $M''F_1^1$, qui ne contiendra plus de variables T, et où la correspondance déterminerea sans ambiguïté où la variable a été omise. Une proposition AXT devra être remplacée par *AX*. Mais à quoi correspondront le *A* et le *X* de la traduction? Nous admettrons que *X* correspond à X (et donc que *AX*, *EX* traduisent AX, EX).
Il s'ensuit que *A* traduira X AXT, donc les prédicats de $M''F_1^1$ seront des prédicats 'relatifs à des cas'.
La correspondance sera (outre la correspondance 1.61) pour

x axt **x** bxt . . . x y
a *b* . . . *x y*

On aura donc *ax*, *bx*, . . . pour axt, bxt.

2.7 POSTULATS ET THÉORÈMES DE $M''F_1^1$

2.71 Si on traduit les postulats 2.1 conformément à 2.6, on obtient un système de postulats équivalent à celui de Barcan (FSI)
2.72 De ces postulats se déduisent les théorèmes de métathéorèmes de Barcan.
2.73 En particulier nous constatons que les théorèmes exclus par Barcan sont les théorèmes mentionnés comme exclus dans 2.26 et 2.27.
2.74 Barcan (IIS) pose des règles d'abstraction pour les abstraits $\hat{x}$ M, de telle sorte que $\hat{x}$ *ax*, expression qu'est la traduction de x axt, constitue l'équivalent de *a*, ce qui est bien conforme à 2.6.

D. EXTENSIONS DE TF_2^1 ET $M''F_1^1$

2.8 EXTENSIONS DE TF_1^1

Nous ne considérons ici que l'extension qui emploierait concurremment des propositions pt_r et des propositions pt.
2.81 Prior semble songer à quelque chose de ce genre lorsqu'il envisage une sorte d'analyse abstractive de 'x possède la propriété a à tel moment' en 'x possède *atuellement* la propriété d'être un a à tel moment'. La 1ère expression (sans abstrait) serait axt, la 2e expression serait (**x** axt) xt_r.
2.82 L'analyse abstractive ci-dessus, sans aboutir à une contradiction, conduit du moins à une sorte de pléonasme vicieux, à l'impossibilité d'appliquer correctement une règle de structure, car (**x** axt) xt_r donnerait quelque chose comme $axtt_r$, ce qui est une expression mal formée.
2.83 Remarquons que la même malformation réapparaît si on tente d'analyser 'x est nécessairement un a' en '*en fait*, x est nécessairement un a'. Et de même pour 'x est possiblement un a'.
Dans le 1er cas on analyserait At axt en (x At axt) xt, donc en (At axt) xt donc (At axt) t. Dans le 2e cas Et axt serait analysé en (**x** Et axt) xt, donc (Et axt) t.
Ceci à moins que le dernier t ne disparaisse dans la réduction, sous prétexte que t est déjà lié dans At axt ou Et axt.

2.9 EXTENSIONS DE M″F

Le caractère malformé de l'expression réduite axtt$_r$ est dissimulé si l'analyse abstractive se fait dans une extension de $M''F^1_1$, où t et t$_r$ ne sont plus explicites.
Instinctivement nous sommes tentés d'analyser *a x* en (*x ax*) *x* comme en logique classique, mais d'effectuer cette analyse en logique modale en interprétant la proposition analysée comme une proposition de logique classique '*x* est *en fait* un $\hat{x}$ *ax*'.
Peut - être est-il possible d'éviter la 'malformation d'expression' par des règles plus ou moins subtiles, jusqu'à preuve de cette possibilité il paraît nécessaire d'écarter l'extension et en particulier le genre d'abstraction en question.

Nous ne traiterons pas en détail de la logique de deuxième ordre $M''F^2$, dont les fondements ont été posés par Barcan (IIS) en vue d'une théorie de l'indentité, celle-ci conduit à une théorie des descriptions étudiées par Carnap (M Ne). A ces diverses parties de $M''F^2$ correspondent des parties de logique TF^2 dont l'étude qui précède pose les fondements.

Université de Louvain

BIBLIOGRAPHIE

Barcan, R. - (FSI) A Functional Calculus od First Order Based on Strict Implication. JSL. XI (1946). (IIS) The Identity of Individuals in a Strict Functional Calculus of the Second order. JSL XII (1947).
Bayart, A. - (CLM) La correction de la logique modale de premier et second Ordre S5. Logique et Analyse I (1958). (QAM) Quasi-adéquation de la logique modale de second ordre S5 et adéquation de la logique modale de premier ordre S5. Ibid II 1959).
Carnap, R. - (M Ne) Meaning and Necessity. Chicago, 1947.
Church, A. - (ATT) A Formulation of the Simple Theory of Types - JSL V (1940).
Feys, R. - (LNM) Les Logiques nouvelles des modalités. Rev. Phil. Louvain 40, 41 (1937–1938). (FMA) Les systèmes formalisés des modalités aristotéliciennes. Ibid. 48 (1950).
Lewis, C. I. and Langford, C. H. - (SyL) Symbolic Logic. New York, 1932.
Prior, A. N. - (FoL) - Formal Logic. Oxford, 1955.
Von Wright, G. H. - (EML) An Essay in Modal Logic. Amsterdam, 1951.

R. FRAÏSSÉ

LES MODÈLES ET L'ALGÈBRE LOGIQUE

I. BREF HISTORIQUE SUR LA NOTION DE MODÈLE EN LOGIQUE MATHÉMATIQUE

Si l'on excepte les tables de valeurs, qui peuvent être considérées comme des modèles pour le calcul des propositions, et qui remonteraient à Frege, la notion de modèle, en logique mathématique, remonte au théorème de Löwenheim-Skolem (1915–1921). Il faut toutefois attendre la *théorie sémantique* de Tarski (Der Wahrheitsbegriff in den formalisierten Sprachen, 1933) pour avoir la définition précise de la valeur – vraie ou fausse – prise par une formule logique pour un modèle. Rappelons qu'on appelle *calcul élémentaire* le calcul logique des prédicats du premier ordre avec égalité. Une *classe élémentaire* (anciennement dite classe arithmétique) est celle des modèles qui vérifient une formule de ce calcul. Deux modèles sont dits *élémentairement équivalents* lorsqu'ils vérifient les mêmes formules.

La théorie sémantique a engendré deux théories qui, chacune, relie la logique mathématique et l'algèbre. La plus connue est la *logique algébrique:* algèbres cylindriques de Tarski et Jonsson, algèbres monadiques et polyadiques de Halmos. On part du calcul logique élémentaire, que l'on veut plonger dans l'algèbre; la notion de formule (proposition ou fonction propositionnelle) reste primordiale, et est algébrisée par abstraction; la notion de modèle est accessoire.

L'autre théorie – fille de la sémantique est l'*algèbre logique*, à laquelle sera consacrée cette communication. On part de l'algèbre dans ses notions les plus élémentaires: relation, isomorphisme, restriction, fonction, homomorphisme, produit direct. Partant de ces bases algébriques, on reconstruit la logique mathematique, et plus précisement certaines, ou même toutes les classes élémentaires. La notion de modèle passe au premier plan alors que la formule devient une notion accessoire.

II. LES PREMIERS PROBLÈMES DE L'ALGÈBRE LOGIQUE

On peut faire remonter l'algèbre logique au théorème de Birkhoff (On the structure of abstract algebras, *Proc. of the Cambridge Phil. Soc., vol. 31, 1935*, p. 433–454): caracterisation de la *classe équationnelle* – définie par une égalité affectée des seuls quanteurs 'quel que soit' – comme une classe de fonctions, fermée pour la restriction, l'homomorphisme et le produit direct. Cette étude a été reprise par Chang (Some general theorems on direct products and their applications to the theory of models, *Proc. Kon. Nederl. Akad., vol 57, 1954*, p. 592–598).
Mentionnons la caractérisation de la classe universelle, ou classe des relations verifiant une formule du calcul élémentaire affectée des seuls quanteurs 'quel que soit', placés au début (voir Tarski, 'Contributions to the theory of models', *Proc. Kon. Nederl. Akad., vol 57, 1954*, p. 572–588, et *vol 58, 1955*, p. 56–64', Vaught, '*Bulletin Amer. Soc.*, vol. 58, 1952, p. 65). Approfondissant la caractérisation de Vaught, j'ai obtenu celle-ci: dire que la classe K est universelle, c'est dire qu'il existe un nombre fini de relations finies $A_1, A_2, \ldots, A_n$ telles que $X \varepsilon K$ lorsque $X \not> A_1$ et $\not> A_2$ et ... et $\not> A_n$ ($B > A$ signifie que A est isomorphe à une restriction de B): voir *Proc. Intern. Cong. of math. Amsterdam 1954*, p. 537–538. On notera par exemple que la classe des relations binaires réflexives est caractérisée par la condition $X \not> F$, où F est la relation définie sur un seul élément a et telle que $F(a, a) =$ faux. La classe des relations transitives est caractérisée par $X \not> A_i$ où chaque A_i est définie sur 3 éléments a, b, c avec $A_i(a, b) = A_i(b, c) =$ vrai et $A_i(a, c) =$ faux, les 6 autres valeurs prises par A_i étant arbitraires (vrai ou faux).
Depuis ces problèmes, bien d'autres questions se sont posées, et d'autres auteurs ont contribué au développement de l'algèbre logique. Citons en particulier Henkin, Ehrenfeucht, et Feferman. Dans la suite, il sera traité de deux problèmes que j'ai particulièrement étudiés, la caractérisation algèbrique des classes élémentaires et l'interprétabilité d'un modèle par un autre.

III. ISOMORPHISME PARTIEL, $\left\{{n \atop p}\right\}$ – ISOMORPHISME, ET $\left\{{n \atop p}\right\}$ – ÉQUIVALENCE

Soit A et A' deux relations à m arguments, définies sur les ensembles

E et E'; une application biunivoque φ de $F \subseteq E$ sur $F' \subseteq E'$est dite un isomorphisme partiel de A vers A' lorsqu'elle est un isomorphisme de A restreinte à F sur A' restreinte à F'. Convenons de dire que tout isomorphisme partiel est un $\left\{\begin{smallmatrix} o \\ p \end{smallmatrix}\right\}$ - isomorphisme quel que soit l'entier naturel p. Que φ est un $\left\{\begin{smallmatrix} n \\ p \end{smallmatrix}\right\}$ - isomorphisme lorsque, quel que soit $\overline{F} = F$ augmenté de $q \leqslant p$ éléments de E, il existe un prolongement $\overline{\varphi}$ de φ à $\overline{F}$ qui est un $\left\{\begin{smallmatrix} n-1 \\ p-q \end{smallmatrix}\right\}$ - isomorphisme de A vers A'; et inversement en remplaçant F, E, φ, A, A', par $F', E', \varphi^{-1}, A', A$. On voit que, si φ est un $\left\{\begin{smallmatrix} n \\ p \end{smallmatrix}\right\}$ - isomorphisme, il en est de même de φ^{-1}, et que le produit de deux $\left\{\begin{smallmatrix} n \\ p \end{smallmatrix}\right\}$ - isomorphismes en est un autre. L'isomorphisme vide (application de l'ensemble vide sur lui-même) est un isomorphisme partiel de A vers A'; lorsque c'est un $\left\{\begin{smallmatrix} n \\ p \end{smallmatrix}\right\}$ - isomorphisme, écrivons $A \overset{n}{\underset{p}{\sim}} A'$; on obtient une équivalence entre relations, et le théorème de caractérisation: dire que $A \overset{n}{\underset{p}{\sim}} A'$ équivaut à dire que A et A' vérifient les mêmes formules du calcul élémentaire, de degré au plus égal à n et ayant au plus p individus liés (on rappelle que le degré d'une formule prénexe est le nombre des alternances de quanteurs 'quel que soit' et 'il existe'). Du théorème précédent on déduit que:

(1) A et A' sont élémentairement équivalentes si et seulement $A \overset{n}{\underset{p}{\sim}} A'$ quels que soient n et p,
(2) une classe K est élémentaire si et seulement si elle est une réunion finie de classes d'équivalences $\overset{n}{\underset{p}{\sim}}$.

(Voir Etude de certains opérateurs dans les classes de relations, définis à partir d'isomorphismes restreints; *Zeitschrift für Math. Logik, vol 2,* no *2, 1956*, p. 59–75; Application au calcul logique *ibid.* p. 76–92.)
En application des notions précédentes, si A, A', B, B' sont des ordres totaux, et si φ est un $\left\{\begin{smallmatrix} n \\ p \end{smallmatrix}\right\}$ - isomorphisme de A vers A', et ψ un autre de B vers B', alors le prolongement commun à φ et ψ défini sur la réunion de leurs domaines, est un $\left\{\begin{smallmatrix} n \\ p \end{smallmatrix}\right\}$ - isomorphisme de $A + B$ vers $A' + B'$. En particulier si $A \overset{n}{\underset{p}{\sim}} A'$ et $B \overset{n}{\underset{p}{\sim}} B'$, alors $A + B \overset{n}{\underset{p}{\sim}} A' + B'$, ce qui entraine immédiatement le théorème de Beth: si A élémentairement équivalent à A', et B à B', alors $A + B$ l'est à $A' + B'$ (Observations

métamathématiques sur les structures simplement ordonnées, *Actes du 2ème Colloque intern. de log. math., Paris 1952*, p. 29–35).

IV. MODÈLE INTERPRÉTABLE PAR UN AUTRE

Rappelons qu'une *théorie* est un ensemble T de formules fermé pour la déduction (si $\varphi \varepsilon T$ et que $\varphi \Rightarrow \psi$, alors $\psi \varepsilon T$) et pour la connection 'et' (si φ, $\psi \varepsilon T$, alors $\varphi \wedge \psi \varepsilon T$). Ou encore T est l'ensemble des formules vérifiées pour certains modèles. Qu'une théorie T est dite *interprétable par S* lorsque chaque axiome de T devient un théorème de S éventuellement complétée par des définitions de prédicats (chaque nouveau prédicat se ramenant à une formule qui ne comprend que les prédicats primitifs, figurant dans les axiomes de S; voir Tarski, Mostowski, Robinson, *Undecidable theories*, 1953, p. 20). L'interprétabilité est réflexive et transitive.
On peut tout d'abord traduire cette notion syntactique en une notion sémantique, celle de *modèle interprétable* par un autre: la relation B est dite *interprétable* par la relation A définie sur le même ensemble, lorsque la théorie complète de B (ensemble des formules vraies pour B) est interprétable par la théorie complète de A. Par exemple si A est un ordre, la relation $B(x, y)$ vraie lorsque x est immédiatement suivi de y, dans l'ordre A, est interprétable par A, au moyen de la formule $A(x, y) \wedge x \neq y \wedge \forall_z (A(z, x) \vee A(y, z))$. L'interprétation inverse n'est pas possible en général: par exemple l'ordre I des entiers naturels n'est pas interprétable par la relation 'suivi de': $S(x, y)$ vraie lorsque $y = x + 1$. Pour le voir, on utilisera le théorème de caractérisation suivant. Dire qui B est interprétable par A équivaut à dire qu'il existe un n et un p tels que tout $\{{n \atop p}\}$ - isomorphisme de A vers A est un isomorphisme partiel de B vers B. Ainsi, pour I et S mentionnés ci-dessus, on peut montrer que, quels que soient n et p, il existe deux entiers u, v pour lesquels la transposition qui échange u et v est un $\{{n \atop p}\}$ - isomorphisme de S vers S. Or ce n'est évidement pas un isomorphisme partiel de I vers I.

V. QUELQUES PROBLÈMES ET APPLICATIONS DE L'ALGÈBRE LOGIQUE

D'abord un problème très précis, mais de portée restreinte. Dire que la théorie T est interprétable par S équivaut-il à dire que pour tout modèle B vérifiant T, il existe un modèle A vérifiant S et tel que B soit interprétable par A?

Deuxième problème: Sur le modèle de ce qu'on a fait pour l'ordre I des entiers naturels, non interprétable par S, on peut espérer obtenir un résultat de la forme suivante: on part de la théorie des ensembles de Bernays-Gödel (axiomes A, B, C, D); Montrer que, pour l'une au moins des relations ε définie sur les classes et vérifiant les dits axiomes, aucune relation binaire interprétable par ε ne peut ordonner totalement la classe des ensembles. Plus particulièrement, montrer que, quels que soient les entiers finis n et p, il existe deux ensembles u et v pour les quels la transposition qui échange u et v est un $\left\{ {n \atop p} \right\}$ – isomorphisme de ε vers ε.

D'un tel résultat, on déduirait que les axiomes A, B, C, D ne peuvent entraîner un axiome assez fort pour définir, avec preuve d'unicité, un bon ordre de la classe de tous les ensembles; en particulier ils ne pourraient entraîner l'axiome de constructibilité de Gödel.

Troisième Problème, beaucoup moins précisé que les précédents. On utilise couramment l'interprétabilité d'une théorie par une autre pour démontrer une consistance relative: si T est interprétable par S et que S soit consistante, il en est de même de T. Ainsi dans les cas simples de la théorie des rationnels interprétable par l'arithmétique des entiers naturels, ou de la géométrie sphérique interprétable par la géométrie euclidienne. Toutefois, dans des cas moins simples mais usuels, l'interprétabilité simple considérée ci-dessus ne suffit pas. On l'approfondit par relativisation des quanteurs (voir Tarski, *Undecidable theories* p. 24); alors seulement on peut, comme l'a fait Gödel, montrer que la théorie A, B, C, D, E est interprétable au sens approfondi par A, B, C, D. Un bonne caractérisation en algèbre logique de cette interprétabilité approfondie est un problème non encore abordé à ma connaissance.

Université d'Alger

MODELS IN APPLIED PROBABILITY

1. It is a pleasure for me that my subject at this colloquium is probability. I think there is no domain of rational activity (daily life excepted) where the use of models is as direct, as palpable, as little problematic as it is in applied probability. So I think my task is much easier than that of anyone who has dealt with models at this colloquium.

Yet I must make some restriction. Probability is the most disputed mathematical notion. It has been the subject of furious discussions. Once in such a discussion, when I had formulated my opinion as clearly as I thought possible and necessary, I learned from one of the answers that I had forgotten the most essential point. I will try to avoid that mistake here.

There are two different kinds of applied probability. One is what *physicists* call statistics, and the other is the statistic of *statisticians*. This difference is already visible in the experimental field. The measuring procedure of the statistician is counting, mostly two sets, such as the number of births, and the number of people in Holland in the same year, and dividing one number by the other (in order to get the so-called birth rate). The statistician proper endeavours to compute probabilities. He does his job by measuring and calculating frequencies and by discussing afterwards the reliability of these frequencies as approximate probabilities. The physicist, however, mostly measures certain magnitudes which are no phenomenological probabilities, but which may be related to probabilities only through a far from simple physical theory. Of course the physicist counting the clicks of a Geiger-Müller-counter behaves as a statistician proper, at least as long as he does not claim to be doing anything more than measuring a decay probability or a half-life of a radioactive material. In the typical cases the physicist applies probability, as he applies Calculus or Fourier transforms, on the base of a non-trivial physical theory. Probability as applied in gas theory is preceded by physical terms such as mechanical system, Hamiltonian, loose interaction, energy, canonical transformations.

On the other hand, the statistician proper applies probability to a material that has not yet been transformed by means of a profound theory. If I remember rightly, it was Laplace who defined probability as common sense pronounced in mathematical language. Indeed probability in statistics is that part of mathematics which applies immediately to nature without any previous physical theory. It is equalled in this respect only by elementary arithmetic. Of course if I predict that two rods of three feet each are six feet when put together, I apply mathematics to real things in the most straightforward way that can be imagined. But applications of probability in statistics are hardly less direct, though the mathematical tools applied in that case are rather profound. We cannot apply potential theory or integral equations to reality without a highly developed intermediate physical theory, working with sophisticated physical models, whereas in statistics proper all specific models can be dispensed with. I will confine myself to this case of applied probability, because the common model of all statistics can be better studied in a case where this model is not tied up with a specific physical model. No scientist is as model-minded as is the statistician; in no other branch of science is the word model as often and as consciously used as in statistics. This is another advantage of choosing probability (as applied in statistics) when dealing with models.

2. No statistician present at this moment will have been in doubt about the meaning of my words when I mentioned the common statistical model. It must be a stochastic device producing random results. Tossing coins or dice or playing at cards are not flexible enough. The most general chance instrument is the urn filled with balls of different colours or with tickets bearing some ciphers or letters. This model is continuously used in our courses as a didactic tool, and in our statistical analyses as a means of translating realistic problems into mathematical ones. In statistical language 'urn model' is a standard expression. Numerous times in courses or lectures I have asked the listeners to imagine an urn containing red, white and blue balls in a given ratio, or black and white ones in an unknown ratio, or an urn containing the waist measurements of all Dutch women, from which I constructed a new urn containing all possible means over 5001 numbers out of the preceding urn, or two urns, one containing white and black balls in the ratio 3:1, and the other containing the same kind of balls in the inverse ratio, with the problem to

guess from which urn some given sample of balls had been drawn. I have never dreamt of bringing with me such an urn, or even of drawing the picture of such an urn on the blackboard, as a physicist would do when dealing with a gas in a vessel or the classical atomic model of hydrogen. Though I may suppose that none of my listeners had ever seen monstrosities like those urns, I am pretty sure that all of them grasp the meaning of my words with more ease than the student of solid geometry who is asked to imagine a dodecahedron or even a cube. Sometimes I add that before drawing a ball, the urn is to be shaken, but even if I forget, I may take it for granted that the student will shake it, as soon as he happens to work with such an urn.

The urn model is to be the expression of three postulates: (1) the constancy of a probability distribution, ensured by the solidity of the vessel, (2) the random character of the choice, ensured by the narrowness of the mouth, which is to prevent visibility of the contents and any consciously selective choice, (3) the independence of successive choices, whenever the drawn balls are put back into the urn. Of course in abstract probability and statistics the word 'choice' can be avoided and all can be done without any reference to such a model. But as soon as the abstract theory is to be applied, random choice plays an essential role.

When we want to define 'random choice' we are committed to reality and realistic models, because no purely formal, that is to say, mathematical definition of random choice has so far been given and because it is extremely improbable that such a definition will ever be found. In statistical practice random choice is made by using a table of random numbers, or so-called Tippett numbers. The first 'Tippett numbers', were produced by means of a stochastic device. So they still reflect the urn model. On the other hand a number sequence produced by a causal machine (a Turing machine) can never be random. In fact it observes the law by which it is defined. It is true that in statistical practice number sequences produced by causal machines can serve as though they were random. As an example we mention the powers a^m mod $10^{10} + 1$ of a natural number a. If the period of this sequence of digits is long enough, they behave in statistical practice as random numbers, because the underlying number-theory law is not likely to be met with in natural material such as a statistician analyzes. But this does not provide any formal definition of random numbers, random choice, and disorder.

It is rather certain that every attempt at defining disorder in a formal way will lead to a contradiction. This does not mean that the notion of disorder is contradictory. It is so, however, as soon as I try to formalize it. In order to say, what is disorder or random choice, I am committed to appeal to some real structure such as the urn model.

3. In 1710 John Arbuthnot, a Scotsman, who was physician-in-ordinary to Her Majesty Queen Anne and the spiritual father of 'John Bull', proved the existence of God using a statistical argument.

According to birth statistics the number of boys among the newborn children in London had surpassed that of girls for 82 consecutive years. If we supposed sex to be a chance event like heads and tails, we could not explain this fact. The probability of a sequence of 82 heads is $2^{-82} \sim 2.10^{-26}$, and this is so small that we are entitled to call it 0 and to say that such an event is impossible. This implies the impossibility of our hypothesis too, that is to say the hypothesis that sex would be distributed over human offspring according to the chance of heads and tails. So the excess of male births proves the intervention of divine providence.

John Arbuthnot's 'proof' is the first known example of mathematical statistics. Of course we do not accept his final conclusion. The sex ratio is not controlled by the chance of head and tail, or at least not by that of a fair unbiased coin that gives the same chance to both sides. But we still learn from Arbuthnot's reasoning that the chance of a male birth is very probably greater than that of a female birth.

Arbuthnot's statistical inference, with its appeal to a model comprising a stochastic device, has become exemplary. In the same way D. Bernoulli and Laplace proved that it cannot be by chance that the inclinations of the planetary orbits against the zodiac are as small as they are found by astronomical evidence. Laplace used this as an argument for his cosmogonic theory. The common aim of those statisticians is a statistical reliability of their judgements of nearly 100%. (At the same time the judgements themselves are rather crude, mostly decisions between some 'yes' or 'no'.) Though in modern statistics we are acquainted with more refined methods, there are still many opportunities to use Arbuthnot's reasoning. Philosophers call it Cournot's principle: if something is proved to be extremely improbable, we are allowed to state that it is impossible. The statement of its impossibility is nearly always stressed by an appeal to something like the urn model. The event to be disproved appears to be as improbable as a large sequence of heads or sixes, when tossing a coin or throwing dice, and so it is impossible.

In modern times statisticians have learnt not to insist on statistical reliability of nearly 100%. They content themselves with a 95% level. (It is not easy to explain why they chose the magic number 5% = 1/20. It seems to be related to the anatomic fact that man has 20 fingers and toes.) This statistical behaviour is motivated by the statistician as follows: Whenever I have to take a decision on the evidence of a random sample, I will choose the procedure of acceptance or rejection in such a way that the probability of a wrong decision will be $< 5\%$ under the most unfavourable conditions. This rule can be expressed by saying that at any moment I will draw my decision out of an urn that contains less than 5% wrong decisions. This means that in the long run the fraction of wrong decisions is not likely to be sensibly greater than 5%, e.g. the probability of a frequency of $\geqq 6\%$ in 65000 decisions will be less than 10^{-30}. According to the Arbuthnot-Cournot rule I may even say that from an urn containing less than 5% wrong decisions it is *impossible* to draw more than 6% wrong decisions in a series of 65000.

4. The problem of statistics is that of inverse probability or probability of causes. I undertake to estimate the contents of an urn on the empirical evidence of a random sample. Though often formulated in this way this problem is impossible. Its reformulation by Thomas Bayes has become classical. We are given not one urn, but a set of urns of known composition (say of white and black balls in a known ratio). A random sample has been drawn from one of them. The problem now runs: knowing the composition of the sample, to guess from which urn it has been drawn (or rather from which ones if there are urns that are not much different in composition). Bayes's procedure can be described as follows: By drawing one ball and ascertaining its colour, I single out a subset of urns from which this ball could have originated. If the ball appears to be white, all purely black urns are to be excluded, all white ones are to be admitted, whereas among intermediate urns of a given composition a fraction is admitted that equals the probability of finding a white ball in such an urn. By every new trial the set of remaining urns is narrowed, and after a long series of n trials, in which p white and q black balls have been found, in the overwhelming majority of remaining urns the ratio of white and black balls will be near $p : q$. If the balls of the

sample have always been drawn from the same urn, this urn will be one of the remaining set, and as this set has become more homogeneous, we can make more precise statements about the interior structure of its urns. Of course we must start with some information about the original set of urns, and this is the weak point of Bayes's theory. Bayes supposes that the composition of the set of urns has been controlled by a chance device, so that I can speak of an urn of urns. Some hypothesis should be made about the so-called a priori distribution of urns, and if this hypothesis is true, Bayes method will improve it step by step. (This is quite another procedure than Arbuthnot's, who makes a hypothesis in order to disprove it). In this century by an ingenious artifice J. Neyman and E. S. Pearson have succeeded in getting rid of the a priori distribution. I must add that as a chance device Bayes does not use white and black balls in an urn. He throws balls upon a rectangular table ABCD, and the things corresponding to white and black balls are balls falling left or right of some parallel to AD that characterizes the corresponding urn; the set of urns is made up by throwing another ball and drawing the parallel to AD through the point where it touched the table. So Bayes supposes an a priori distribution of urns in which every composition of urns occurs with the same probability. Of course Bayes's chance device does not differ essentially from the urn model.

5. The urn model reappears when I try to explain statistical methods in the most modern version, Wald's theory.

I suppose that there are two kinds of urn, one kind containing 3/4 white and 1/4 black balls, and the other kind showing the inverse ratio. The first kind will be called white urns, and the second kind black urns, though the exterior of both will be the same.

I am offered some urn for \$ 100. I would like it to be a white one. If it is really so, I can sell it with a gain of 10%. If I am mistaken, I have bought a worthless thing, and the \$ 100 paid for it will be lost.

I am allowed to examine the quality of the urn, but I shall restrict myself to the inspection of a sample that I shall draw at random. The cost of inspection will be \$ 0.10 a ball. The sample should be fairly small, otherwise the cost of inspection would consume the expected gain. Let us try a sample size $n = 10$.

The chance of finding u white balls in a sample of 10 drawn from a white urn, can be calculated by the binomial formula. It is

$u =$	0 1 2	3	4	5	6	7	8	9	10
	0.000	0.003	0.016	0.058	0.146	0.250	0.286	0.188	0.056.

For a black urn the figures will be the same, but in the inverse order. After having viewed the sample I will decide whether I shall buy the urn or not. The decision shall depend on a rule of acceptance stipulated in advance. An appropriate rule seems to be (R^{10}_{6}):
≤ 6 white balls in the sample $\rightarrow$ call the urn black; do not buy;
> 6 white balls in the sample $\rightarrow$ call the urn white; buy.
Then the following table specifies the probability of a

		black urn	white urn
to be called	black	99.7 %	22.0 %
	white	0.3 %	78.0 %

There are two kinds of error I can commit: (1) to call a black urn white, (2) to call a white urn black.
In the first case I shall lose \$ 100 because I shall have got a worthless thing. In the second case I shall not buy the urn, though I should have gained \$ 10 if I had done so. In either case the expenses for the sample inspection are booked on the loss account. Clearly errors of the first kind are more dangerous than errors of the second kind. For this reason I picked out a rule that leads to a small probability in the left lower corner of the table.
I shall calculate the gain expectation of this transaction. The sprat I have thrown out is the \$ 1 for the sample inspection. The whale I will catch [1]), is the gain of \$ 10 on the sale of the urn. But I shall not catch the whole whale. If the urn is black, I do not catch it at all, or rather there is no whale. But if the urn is black the chance that I walk into the trap is as small as 0.003. The expected loss in this case is $\$ 100 \times 0.003 = \$ 0.30$ (increased by the cost of the sample). If the urn is white, there is a chance of 0.78 that my rule will detect its whiteness and that I shall buy it. So the expected gain will be $\$ 10 \times 0.78 = \$ 7.80$ (and the cost of the sample).
I am faced with a gain expectation of \$ 6.80 and a loss expectation of

[1]) 'The sprat to catch a whale' is not only a saying, it is a model. So are 'the needle in the bundle of hay', 'Augias' stable', 'the Danaid's vessel', 'tantalizing', and many others. It would be interesting to study models in daily life.

\$ 1.30. Is the one sufficiently counterbalanced by the other? The answer on this question will depend upon the probability that urns might be faked. If the probability is big, I will not risk a loss of \$ 1.30 for a gain of \$ 6.80. If it is small, I will. I need some information about the probability of urns being white. This information may be rather vague. I apply essentially Bayes's procedure: improving on a priori information by sampling. No transaction is possible as long as I have no idea about that a priori probability. I shall not risk the \$ 1 for the sample inspection if there is so much trickery that the chance of an urn being white is nearly nought. I will even refuse to admit the seller, if I judge that chance to be very low.

In our case the a priori probability of an urn being white, should be at least 13/81 if I hope to get off with a whole skin. The expected gain will then be

$$\$\,6.80 \times \tfrac{13}{81} - 1.30 \times \tfrac{68}{81} = 0.$$

Our computations depend on the chosen rule of acceptance. I may single out another, and I may compare rules and try to find the best one under a certain set of a priori distributions. The details of that procedure will be of minor importance for our problem.

6. Probability theory has arisen from games. For two centuries statistical problems have been analyzed according to the model of an unselfish game. It has been the aim of the statistician to make some statement about the content of an urn. Under the same condition this could be a rough statement with a large reliability, or a more refined, but less reliable one. There are a great many statistical behaviour rules. No choice is possible as long as I do not account for their individual consequences. This obvious truth has been disguised (and it is still so) by the acceptance of conventional rules with respect to the level of reliability of statistical decisions, such as the 95% rule of 'fingers and toes statistics'. In fact any good statistician will account for the consequences of wrong or right decisions. In Wald's theory this is done systematically. Wrong and right decisions are to correspond to numerical losses and gains. In quality control and drug testing it is not difficult to settle the gain and loss table. But even in scientific work this is not as impossible as it seems. A statement like 'the elementary charge is $(4.803 + 0.005)\,10^{-11}$ E.S.U.' aims at a statistical decision. It tells me the reliability with which I can assert the

elementary charge to be comprised between two bounds. As long as its value is an abstract magnitude, no statistical problem will arise. But as soon as my behaviour is essentially different according as that value is $\geq 4.85 \times 10^{-11}$ or $\leq 4.75 \times 10^{-11}$, I must be able to estimate numerically the losses or gains caused by a wrong or right decision.

In order to exclude unselfish games, I will speak of a bet, if I allude to Wald's statistical model. In applied probability, especially in statistics, the bet model may be expected to supersede the pure urn model, though at present practical statisticians are not yet sufficiently bet model minded. In statistical evaluations of scientific results and generally in all cases where the gain and loss table is not evident, the role of the bet model is not yet clear enough. I think that the situation has been obscured by Wald's saying that in those cases the work of the statistician is a struggle against nature. I think this is not more than a witticism that does not contribute in any way to the solution of the problem. I am sure that the bet model is correct in those cases too, but I do not believe in an interpretation where Nature should be assigned as betting partner.

In our context the bet model is not as fundamental as the urn model. Once the urn model has been accepted, the bet model is no more than a refinement. I have stressed the informal character of the urn model: it cannot be replaced by a formal, purely mathematical device. It represents the appeal to reality that is needed as soon as mathematical probability is to be applied. Bets, however, can formally be defined, on the basis of the urn model. It is true that a bet includes an agreement about some payments that should be performed if a ball of some colour is drawn out of the urn. But in order to understand the bet, we need not be told what the word 'payment' means. The only thing that matters is the minimax behaviour rule that can be pronounced in purely mathematical terms. We must be told that every bet partner will try to maximize his gain (or to minimize his loss) under the most unfavourable conditions. So after the urn model has been accepted, there is no room for new informal devices in order to establish the notion of a bet.

7. I am sure you will have got the impression from my exposition that the urn model is indispensable if mathematical probability has to be applied. This was my own conviction until, a few years ago, I realised that this is not correct. There is a quite different model that can serve as well as the urn model. I noticed this when I tried to elaborate a language

in which we might communicate with humanlike beings not acquainted with any one of our natural languages, not even with our mathematical mode of expression. This lecture is not the place to give details of that project. It will suffice to mention in this connection that so far four chapters have been developed dealing with respectively: 'Mathematics', 'Time', 'Behaviour' and 'Space, Motion, Mass'. It is a noteworthy fact that many features of human behaviour could be shown and named before the introduction of any notion from the material world, any notion of mechanics. At a certain point in the chapter 'Behaviour' I was struck by the possibility of introducing probability, not as an abstract mathematical notion, but in its applied form. This could be done without any appeal to the material world, at a moment when there was not the slightest opportunity of speaking about urns that could be shaken, coins that could be tossed, or other material devices whatsoever.

I introduced two actors playing some game like matching pennies. In its usual form this game is played by two persons A and B who simultaneously show a penny. If the shown result is two heads or two tails, A wins the bet; if it is one head and one tail, B wins.

Owing to Von Neumann's work the theory of games has become a favourite topic of research nowadays, but even if you have never heard of this theory, you will have no difficulty in discovering the essential features of that game. Of course, when playing it, you will try to increase your chance, but the only way to do it is to discover regularities in the actions of your opponent, and to avoid such regularities in your own behaviour. As soon as one of the players falls into some regularity, the other player can take advantage of it. But there is one minimax strategy for both players, that is to say a strategy that is the best under the worst circumstances. The minimax strategy consists in the random choice of head and tail of the coin, both sides with the same probability 1/2. In order to realize this strategy, players may be advised to avail themselves of some stochastic device, like an unbiased coin that tells which side of the coin should be shown. Otherwise it would not be easy not to betray oneself. People who are asked to reel off numbers will quickly fall into repetitions. A skilled player can read them like a book. So does Shannon's machine that scores 60% when playing matching pennies against human partners.

In Shannon's and my own version matching pennies is played by choosing simultaneously one of the numbers 1 and 2. If both choose 1 or both choose 2, A wins, otherwise B wins. Any humanlike intelligent being that understands human behaviour and the rule of this game will discover the minimax strategy, especially if illustrated by a few examples. The complete strategist playing against another complete strategist is a minimax strategist, and the minimax strategist in matching pennies is equivalent to an urn producing two events with equal probabilities. This means that in all applications such an urn can be replaced by a human, a complete strategist playing matching pennies against another complete strategist. It is evident that this is not a special feature of urns containing white and black balls in the same ratio; it is easy to find out a game that corresponds to a given urn, so that the complete strategist of this game considered as a stochastic device is equivalent to the given urn.

It does not matter by which means the complete strategist actually ensures his completeness. Eventually this might be done by using a mechanical chance device, but in a definition of the complete strategist the special means of realization need not be unveiled. So you will understand that in the earlier mentioned context it was possible to introduce probability and probabilistic notions like random choice at a stage when no appeal to material reality was possible. Instead we used what are essentially features of human behaviour.

I think it is a noteworthy result that the seemingly indispensable urn model can be dispensed with in applied probability. In the bet model the material chance device can be replaced by a complete strategist playing against another complete strategist. Of course, the model of the complete strategist equivalent to an urn is as little formal as is the urn. It is not a mathematical device. In order to define it we must appeal to reality, though not to the material reality of an urn, but to that of human behaviour.

After all, this is not to be wondered at. Historically, probability and probabilistic notions like disorder and random events have their roots in such important utterances of human behaviour as games. In this respect disorder and chance do not differ from order and law which are not less deeply rooted in human behaviour such as it is unveiled in human society.

University of Utrecht

G. FREY

SYMBOLISCHE UND IKONISCHE MODELLE

(1) Wir verwenden Modelle in verschiedenem Sinne und zu ganz verschiedenen Zwecken, wie das L. Apostel darlegte. Ein Modell ist wohl immer aufzufassen als eine Abbildung. Die Frage ist nur, was abgebildet wird, und wie die Abbildungsfunktion aussieht.

'Modell' aus dem Lateinischen modus, modulus abgeleitet, hat seine ursprüngliche Bedeutung in der Baukunst. Von Vitruv übernommen ist es schon im Mittelalter ein Maß (Halber Säulendurchmesser), in dem alle Verhältnisse eines Bauwerkes ausgedrückt werden. Und so entstehen in den europäischen Sprachen die Worte moule (afz.), mould (aengl.), model (ahd.), die alle schon die Bedeutung von Form haben, etwas nach dem 'rechten Maß' Gemachten. Im 16. Jahrhundert wird das italienische modello = Muster, Vorbild (Proplasma) neu übernommen als modèle (fz.), model (egl.), Modell (d.). Hier hat es den in der bildenden Kunst gebräuchlichen Sinn einer Musterform, nach der etwas gemacht wird. Insofern dieses meistens verkleinert ist, bürgert sich die Vorstellung ein, Modell 'heisset überhaupt eine jedwede cörperliche Abbildung eines Dinges ins kleine, oder ein nach verjüngtem Maß-Stab verfertigte und einem größeren Cörper ähnlich gemachten kleineren Cörper ...' (Zedlers Universallexicon, 1739).

Die in Kunst und Kunsthandwerk verwendeten Modelle sind im allgemeinen aus einem anderen Material verfertigt als das Original. So sind bereits im 16. Jahrhundert bei Goldschmieden italienische modelli gebräuchlich, die aus Gips oder Blei hergestellt waren. Bereits in dieser ursprünglichen Verwendung des Modellbegriffes zeigt sich also, dass es sich um eine partielle Abbildungsfunktion handelt, die nur die äußeren Formen geometrisch ähnlich abbildet.

Der Modellbegriff bezieht sich als ein maßstabsgerechtes, ähnliches, im allgemeinen verkleinertes Abbild ohne Zweifel auf eine bestimmte vorgegebene Objektgesamtheit, ein 'materielles System'. Wenn wir auch im weiteren die Beschränkung auf eine maßstabsgerechte, geometrisch ähnliche Abbildung aufgeben, so bleibt doch die Abbildung eines bestimmten, vorgegebenen Objektbereiches.

Die Anwendung des Modellbegriffes in der Naturwissenschaft des 19. Jahrhunderts gibt diesem eine neue Bedeutung. Bereits im Altertum gibt es in der Gegenüberstellung von mathematischer und physischer Astronomie die Auffassung, daß die mathematischen Hypothesen nur zur Berechnung der Himmelsvorgänge da sind, aber nichts über deren physische Beschaffenheit aussagen. Auch die ptolemäischen Vorstellungen des Almagest stellen in diesem Sinne nur ein Modell dar. Die nachkantische Philosophie und Naturwissenschaft kann sich der Konsequenz immer weniger entziehen, daß die Hypothesen, die theoretischen Entwürfe der Naturwissenschaften nicht eine an-sich-seiende Wirklichkeit wiedergeben. Sie sind nur Modelle. Die Auffassungen der einzelnen Vertreter unterscheiden sich nur darin, ob angenommen wird, daß es eine solche an-sich-seiende Wirklichkeit überhaupt gibt und wenn dies der Fall ist, ob sie nicht vielleicht unerkennbar sei. In diesem Sinne spricht etwa M. Planck von dem Modell, welches wir das 'physikalische Weltbild' nennen. Dieses ist für ihn eine 'modellmäßige Idealisierung'. Wesentlich ist für ihn dabei, daß alle Zahlwerte von Maßgrößen als unmittelbare Messungen mit einer Unbestimmtheit behaftet sind, im Modell dagegen handelt es sich um ideale Größen, die ganz bestimmte Zahlwerte haben. Das Modell ist also etwas, mit dem wir anstelle einer nicht faßbaren Wirklichkeit operieren. Insofern das Modell auch in diesem modernen Sinn noch Abbildungsfunktion hat, bezieht es sich auf die Beobachtungen und Messungen.

In einem etwas anderen Sinne wird der Modellbegriff mehr in ursprünglicher Bedeutung verwendet als Abbildung eines Erscheinungsbereiches auf ein anderes genauer bekanntes und leichter verständliches. In diesem Sinne wurde sowohl von Ch. Huygens als auch von I. Newton versucht, die optischen Erscheinungen auf Erscheinungen der Mechanik abzubilden, also ein mechanisches Modell der Optik zu geben. Auch hier tritt uns wieder die Tatsache entgegen, daß die Modellbeziehung eine partielle Abbildungbeziehung darstellt. A. Maxwell hat die Partialität dieser Abbildungsbeziehung auch 'physikalische Analogie' genannt. 'Unter einer physikalischen Analogie verstehe ich jene teilweise Ähnlichkeit zwischen den Gesetzen eines Erscheinungsgebietes mit denen eines anderen, welche bewirkt, daß jede das andere illustriert.' Der von H. Hertz in den Prinzipien der Mechanik definierte Modellbegriff zielt in die gleiche Richtung.

Wir wollen den Modellbegriff im weiteren in zweifacher Bedeutung verwenden. Modell ist einerseits zu unterscheiden von der reinen Theorie. Deren wesentliches Charakteristikum können wir mit Cl. Maxwell darin sehen, daß ihre Aussagen keine speziellen Zeit- und Ortskoordinaten enthalten, sie sind Allsätze gerade bezüglich der Zeit- und Ortskoordinaten. Demgegenüber stellt das Modell in dieser ersten Bedeutung die Abbildung eines bestimmten Objektsystems dar, das gerade als solches durch spezielle Zeit- und Ortskoordinaten charakterisiert ist. Dabei kann sich die Partialität der Abbildungsfunktion durchaus auch gerade auf die speziellen Koordinaten beziehen, z.B. in einem topologischen Modell. Der Modellbegriff kann sich aber auch unmittelbar auf die Theorie beziehen. In diesem Sinne sprechen wir von Atommodellen und ähnlichem, hierher gehörten aber wohl auch die Auffassungen Plancks, Huygens', Newton's und Hertz'.

(2) Die Abbildung eines Objektbereiches auf ein anderes stellt eine Zuordnungsfunktion dar. Die Stationen des Eisenbahnnetzes eines Landes können eineindeutig abgebildet werden auf eine Karte dieses Eisenbahnnetzes. Jeder Station entspricht ein Punkt, jede Strecke zwischen zwei Stationen einer Linie auf der Karte. Die Karte ist bezüglich der Nachbarschaftsrelation der Stationen eine isomorphe Abbildung des Eisenbahnnetzes der Wirklichkeit. Durch eine Modelleisenbahn ließe sich außer dem Netz der Nachbarschaftsrelationen auch etwas von der Funktion der Eisenbahn abbilden.

In all solchen Fällen wird niemals alles aus der Wirklichkeit abgebildet. Wir abstrahieren und betrachten nur die Übereinstimmung einiger wohldefinierter Eigenschaften. Die Netzkarte ist nur ein Modell bezüglich der Nachbarschaftsrelation. Eine maßstabsgerechte Karte ist ein Modell bezüglich aller Entfernungen und Winkel. Der Fahrplan stellt ein Modell des Fahrbetriebes dar, das sich bloß auf die Weg-Zeit-Funktion bezieht, wobei es gleichgültig ist, ob man einen graphischen Fahrplan oder ein Kursbuch benützt.

Karte und Fahrplan haben eine praktische Bedeutung. Mit ihrer Hilfe kann ich mich der 'Wirklichkeit', d.h. der Eisenbahn bedienen, um von einem Ort A zur rechten Zeit nach B zu gelangen. – Wir entnehmen dem Beispiel, daß der Begriff des Modells eine partielle Abbildungsbeziehung auf ein gegebenes Objektbereich beinhaltet.

Die Partialität der Abbildungsbeziehung schließt selbstverständlich den

Fall der vollständigen Isomorphie als Sonderfall mit ein. Der Charakter der Partialität ist von Fall zu Fall sehr verschieden und daher nur schwer allgemein zu formulieren. Wir können uns z.B. denken, daß das Objektsystem S eine endliche oder unendliche Menge A von Beziehungen enthält, so daß A alle Beziehungen von S enthält. Das Modellsystem M enthält eine Menge B von Beziehungen, so daß gilt $B \subset A$. Wir können auch sagen, ein Teilsystem von S ist isomorph zu M. Unter Umständen wird man wohl auch den Fall betrachten müssen, daß ein Teilsystem von S einem Teilsystem von M isomorph ist. Der erstere Fall trifft auf das Beispiel der topologischen Netzkarte des Eisenbahnnetzes zu. Aber auch der Modellbegriff, den H. Hertz in den Prinzipien der Mechanik definiert hat, fällt darunter. Zwei Modelle sind danach Modelle voneinander, wenn alle Zeit- und Ortskoordinaten eindeutig und maßstabsgerecht aufeinander abgebildet werden können, abgesehen wird insbesondere von den Masseneigenschaften. Die Isomorphie besteht also nur zwischen einer Teilmenge von Beziehungen.

Es soll nur noch darauf hingewiesen werden, daß auch noch ein anderer Fall möglich ist: Es können gewisse Beziehungen im strengen Sinne nicht in S enthalten sein. Es werden dann jeweils Teilmengen von Beziehungen von S abgebildet durch entsprechende Grenzwertbeziehungen in M (z.B. ein System 'starrer Körper' als Modell eines Systems deformierbarer Körper).

Wir haben des weiteren zu unterscheiden, auf welche Art von Objektbereichen sich die Abbildungsbeziehung bezieht. Wir können nun die oben gemachte Unterscheidung zweier hier zu betrachtender Modellbegriffe durch die Abbildungsfunktion definieren. Modell im ersteren Sinne soll sich immer auf ein materiell-empirisches Objektbereich beziehen. Die abgebildeten materiell-empirischen Objektbeziehungen sind dabei nicht einfach materielle Dingbeziehungen im Sinne an-sich-seiender Wirklichkeit, sondern nur Beziehungen zwischen Handlungen, Operationen und Beobachtungen, die mit und an jenen Objekten gemacht werden. Die physischen Objekte sind uns nur gegeben entweder

(1) durch die Summe ihrer sinnlichen Wahrnehmungen, ihrer Erscheinungen (phänomenologische Auffassung),

oder

(2) durch die Summe der Handlungen und Operationen, die man mit ihm ausführen kann (operative Auffassung),

oder

(3) durch die Phänomene *und* Operationen (allgemein konstruktive Auffassung).

Ganz gleich auf welchen Standpunkt wir uns stellen, auf alle Fälle sind die experimentell oder durch Beobachtung festgestellten Beziehungen nur Beziehungen zwischen Erscheinungen, Phänomenen, Handlungen und Operationen. Nur diese werden in unseren Modellen abgebildet.

Modell im zweiten Sinne ist eine Abbildung eines theoretischen Systems auf ein anderes, es ist also eigentlich ein Modell eines Modells. Es wird zur Illustrierung im Sinne Maxwells aber auch dazu verwendet werden, um schwer verständliche Beziehungen vereinfacht darzustellen und sie leichter anwendbar zu machen. So ist etwa die von den theoretischen Chemikern verwendete Theorie der Mesomerie ein solches Modell der exakteren und viel schwieriger zu handhabenden quantentheoretischen Beziehungen.

(3) Ein physikalisches Modell bildet partiell ab, ist also eine physikalische Analogie. Der Analogiecharakter kommt, wie wir oben schon sagten darin zum Ausdruck, daß das eine das andere Erscheinungsbereich illustriert.

Dieser 'illustrierende' Charakter eines Modells wird immer vorliegen, wenn dieses nur bereits bekannte Vorstellungen verwendet. So ist zumindest bis zum Ende des vorigen Jahrhunderts die vorherrschende Idee, alle Erscheinungen durch mechanische Modelle zu erklären. In diesem Sinne versuchte Maxwell die Experimente, die Faraday durch die Vorstellung der Kraftlinien erklärte, durch das Modell einer inkompressiblen Flüssigkeit darzustellen. Es gibt keine inkompressiblen Flüssigkeiten. Trotzdem sehen wir keine Schwierigkeit darin, uns solche vorzustellen, genauso wie wir uns ohne weiteres starre Körper denken, obwohl wir wissen, daß es solche nicht gibt. Diese Modelle sind also selbst Abstraktionen der Wirklichkeit. Maxwell geht bei der Entwicklung seiner Theorie der Faraday'schen Kraftlinien sogar noch einen Schritt weiter, indem er für seine hypothetische Flüssigkeit auch noch positive und negative Quellen, Entstehungs- und Vernichtungsstellen einführt. Die Freiheit, seiner Modellflüssigkeit beliebige Eigenschaften zu geben, wenn es zur modellmäßigen Darstellung der elektromagnetischen Erscheinungen erforderlich ist, betont er ausdrücklich: 'Gerade so, wie es uns freistand, sie absolut unzusammendrückbar vorzustellen, können wir jetzt voraussetzen, daß sie an gewissen Stellen aus nichts hervorgebracht und an anderen wieder in nichts aufgelöst wird.' Die Bildhaftigkeit der Modellvorstellung wird gewahrt, solange es sich um Vorstellungen handelt, die nur durch gewisse räumlich-zeitliche Prozesse darstellbar sind. In diesem Fall wollen wir ein Modell als *ikonisch* bezeichnen.

Nach Ch. W. Morris ist ein Zeichen ikonisch, wenn es dem, was es bezeichnet, in einem anschaulich-bildlichen Sinne ähnlich sieht, wenn es dem Betrachter sofort kundgibt, was es bezeichnet. Zeichen, die nicht in dieser Weise 'erkennbar' sind, werden als nichtikonisch bezeichnet. Es ist klar, daß die Frage, ob ein Zeichen ikonisch ist oder nicht, nicht immer eindeutig entscheidbar ist. Stilisierte Zeichen können durchaus einen erkennbaren ikonischen Bezug auf das Bezeichnete haben, trotzdem aber nicht für jedermann erkennbar sein. So sind die Zeichen einer Schaltskizze sehr wohl ikonisch, d.h. sie geben bildhaft etwas wieder, sind aber doch nicht ohne weiteres für jeden verständlich. Es gehört eine Vorkenntnis dazu. Man muß in vielen Fällen erst auf die Ähnlichkeit des Zeichens mit dem Bezeichneten hingewiesen werden. Wir wollen ein Zeichen als ikonisch bezeichnen, wenn eine solche Ähnlichkeit überhaupt vorhanden ist. In andern Falle heiße das Zeichen nichtikonisch.

Entsprechend dieser Unterscheidung hat M. Bense vorgeschlagen, zwischen ikonischen und nichtikonischen Modellen in der Physik zu unterscheiden. Das ikonische Modell als ganzes soll also einen anschaulich-bildhaften Bezug auf das Abgebildete haben. Das ist immer der Fall, wenn es sich um eine bloße Abstraktion des Abgebildeten handelt. Der Zusammenhang mit der Morris'schen Zeichentheorie liegt hier auf der Hand. Die Schaltskizze ist aus ikonischen Einzelzeichen aufgebaut, deren syntaktische Verknüpfung selbst wieder ikonisch ist. Wir nennen dann ein solches zusammengesetztes, strukturiertes Zeichen auch Modell. Es ist eine unmittelbare Abbildung einer speziellen Wirklichkeitsstruktur, wir wollen es daher auch als *primäres ikonisches Modell* bezeichnen und damit andeuten, daß es sich um ein Modell in der ersten Bedeutung handelt.

Stellt die Schaltskizze einen Schwingkreis dar, so kann der Physiker sofort eine weitere Abbildung angeben, indem er die Gleichungen eines solches Schwingkreises aufstellt [1]). Die Gleichungen stellen wieder ein

[1]) Das Beispiel wurde von H. Freudenthal in der Diskussion aufgestellt und diskutiert.

Modell der gleichen speziellen Wirklichkeitsstruktur dar. Es besteht keine unmittelbare bildhafte Ähnlichkeit mit dem Abgebildeten, das Gleichungssystem stellt ein *symbolisches Modell* dar. Wie sieht nun die Beziehung zwischen einem primär ikonischen und einem symbolischen Modell aus, die einander abbilden, bzw. beide die gleiche spezielle Wirklichkeitsstruktur abbilden?

Eine geometrische Raumkurve möge das ikonische Modell der Bahn eines bewegten Körpers sein. Die Raumkurve lasse sich hinreichend genau durch eine Gleichung beschreiben, die somit als ein symbolisches Modell der Bahn des gleichen Körpers angesehen werden kann. Um von der geometrischen Raumkurve zur Gleichung übergehen zu können, müssen wir zunächst Arithmetik und Algebra kennen. Wir müssen also das zur Abbildung verwendete symbolische System kennen, d.h. insbesondere seinen Zeichenvorrat (Semiotik), seine Syntax und seine Regeln. Hinzu kommen die speziellen Abbildungsbeziehungen, die in diesem mathematischen Beispiel als reine Mengenzuordnungen aufzufassen sind.

Analog steht es in dem oben gegebenen Beispiel der Schaltskizze eines Schwingkreises. Neben Arithmetik und Algebra treten hier die Gesetzmäßigkeiten der Elektrodynamik. Die Definitionen der verwendeten Größen (Widerstand, Kapazität, Induktivität etc.) stellen die Zuordnungsbeziehungen dar. Wir müssen also auch hier das verwendete symbolische System und die Abbildungsbeziehungen kennen. Diese könnten wir auch als die Semantik des symbolischen Modells bezeichnen.

Auch die Betrachtung anderer Beispiele zeigt, daß es sich um eine gegenseitige Korrespondenz von ikonischem und symbolischem Modell handelt. Wir kennen die physikalischen Gesetzmäßigkeiten in einer symbolisch-mathematischen Form und suchen nun zu diesen Veranschaulichungen, indem wir uns von ihnen ikonische Modelle machen. Handelt sich dabei wieder um eine Anwendung der Theorie (in symbolischer Form) auf eine spezielle Wirklichkeitsstruktur, etwa eine vorliegende Versuchsanordnung, so erhalten wir wieder ein primär ikonisches Modell. Diesem entspricht also in der symbolischen Form die gesamte Theorie ergänzt durch die speziellen Randbedingungen.

(4) Auch die Theorie kann als ein Modell der Wirklichkeit angesehen werden (vgl. die oben zitierte Auffassung Plancks). In ihr werden wieder

die Gesetzmäßigkeiten in symbolisch-mathematischer Form dargestellt. In der klassischen Physik gehen alle Theorien auf ikonische Modelle zurück. Daß die klassische Mechanik an einem solchen ikonischen Modell als Abstraktion der Wirklichkeit abgeleitet wurde, ist unmittelbar verständlich. Aber auch das oben erwähnte Beispiel der Herleitung der Theorie des Elektromagnetismus durch Maxwell zeigt, daß ein ikonisches Modell dieser Herleitung zugrunde lag. Wir wollen in all diesen Fällen von einem *sekundär ikonischen Modell* sprechen, und damit auch zum Ausdruck bringen, daß es sich dabei um Modelle in der zweiten Bedeutung handelt.

Auch in all diesen Fällen besteht eine Korrespondenz zwischen ikonischem und symbolischem Modell. Das symbolische Modell kann als gegeben aufgefaßt werden und dann läßt sich zu ihm ein entsprechendes abbildendes ikonisches Modell konstruieren. Nicht immer braucht eine solche Veranschaulichung identisch zu sein mit dem ikonischen Ausgangsmodell. So wird heute kaum noch ein Physiker das Maxwell'sche ikonische Modell einer inkompressiblen Flüssigkeit verwenden. Statt dessen wurde der ebenfalls ikonische Modellbegriff des Feldes entwickelt. Wir wollen ein symbolisches Modell, zu dem es ein vollständiges abbildendes ikonisches Modell gibt, als *ikonisch-symbolisches Modell* bezeichnen. Es besteht nun die Möglichkeit, daß die symbolisch-mathematisch formulierten Gesetzmäßigkeiten sich nicht durch eine Theorie darstellen lassen, die durch ein ikonisches Modell vollständig abbildbar ist. Wir haben dann ein *nichtikonisch-symbolisches Modell* vor uns. Die Heisenberg'sche Matrizenmechanik ist hierfür ein Beispiel ebenso wie die ihr gleichwertige Schrödinger'sche Wellenmechanik.

Das symbolische Modell ist in diesem Falle nichtikonisch, weil es nicht vollständig auf ein ikonisches Modell isomorph abbildbar ist. Das schließt nicht die Möglichkeit aus, daß es partielle ikonische Modelle gibt. Diese muß es vielmehr immer geben. So ist die Dirac'sche Theorie des Elektrons in unserem Sinne ein nichtikonisch-symbolisches Modell. Hönl hat ein Modell des Elektrons entwickelt, das zwar die Dirac'sche Theorie nur partiell wiedergibt, aber dafür ikonisch-symbolisch ist.

Technische Hochschule, Stuttgart

LITERATUR

Grimm, Deutsches Wörterbuch, 6 Bd. 1885.
M. Planck, Die Kausalität in der Natur, in: *Wege zur physikalischen Erkenntnis*, Leipzig 1934.
J. Cl. Maxwell, Über Faraday's Kraftlinien, *Transact. of the Cambr. Phil. Soc.* Vol. 10, Dtsch. v. L. Boltzmann, Leipzig o. J.
H. Hertz, Die Prinzipien der Mechanik, Leipzig 1894.
G. Frey, Gesetz u. Entwicklung in der Natur, Hamburg 1958.
G. Frey, Ein Satz betreffend das Verhältnis empirischer und symbolischer Objektbereiche, *Philosophia Naturalis*, IV, 2/3, 1957.
Ch. W. Morris, Foundations of the Theory of Signs, Chicago 1938.
Ch. W. Morris, Signs, Language and Behavior, New York, 1946.
M. Bense, Der Begriff der Naturphilosophie, Stuttgart 1953.
H. Hönl, *Z.S.f.Naturforschung* 1, 1946.
H. Hönl, *Math.-physik. Semesterber.* Bd. 1, 1949.

THE MODEL IN PHYSICS

It must have been in an unthinking moment that I promised to speak about 'the' model in physics. For even in physics models of quite different types are used for quite different purposes. Rather than talk about 'the' model, I will take a few examples of some types of models in physics. The catalogue of types will by no means be exhaustive.
Two extreme types are the model as a more or less exemplary ideal and the model as a more or less poor substitute. In various types the ideal meaning and the substitute meaning appear intermixed. A sharp classification of types seems unworkable. And I know of no model of a model.

I. REPRESENTATIVE MODEL

Representative models may be built for demonstration or for experimental investigation. In a planetarium the planetary motion is imitated by a mechanism with entirely different equations of motion. In 1780 such a planetarium was built in Franeker by Eise Eisinga in order to demonstrate that a planetary conjunction would not result in a catastrophic collision. More direct are small scale models, e.g. hydrodynamic models of ships or tidal currents, for manageable experimental investigation. The equations of motion and the boundary conditions are roughly the same as in the original system, but some details may be changed by the scale transformation, e.g. the tidal current deviation by Coriolis forces. For our present topic a still more interesting type of representative model is the analogue model. Its equations of motion are, apart from scale constants, formally similar to those of the original system, but the mechanism may be entirely different. In this way e.g. the motion of single electrons in an electronic tube may be imitated with small balls rolling on a rubber sheet. The sheet is braced over models of electrodes at heights proportional to the potentials of their originals. In a similar way the motion of radio signals in the ionosphere may be imitated with balls rolling on a curved surface. The height of the surface is now proportional to the corresponding electron density and the initial velocity of the balls is proportional to the radio frequency. The analogy is inter alia impaired because the balls are subjected to mechanical friction and do not show diffraction effects. Analogue models of this type are rather specialized. Far more flexible and abstract are electronic analogue computers, which may be adapted to very different kinds of systems. The search for the most perfect computer would lead us back to the most specialized model, viz. the original system itself. But that would entirely overshoot the mark. It is an essential feature of representative models that parameters or experimental conditions may be varied more easily or cheaply than in the original system. Moreover the model may be used for testing theoretical suppositions about the system.

II. SUBSTITUTE MODEL

Provisional simplified models often occur as conscious theoretical approximations. Usually they have a more or less intuitive or illustrative pictorial character. They serve as an intermediate stage between an observed phenomenon and the more fundamental theory by which it should actually be treated. Maybe the application of the fundamental theory would be too complicated, maybe the fundamental theory has been developed insufficiently or not at all.

The old quantum theory underlying the atomic model of Niels Bohr was still incomplete. Later it became completed in a certain sense into the more or less closed formalism of quantum mechanics. Atomic systems now fall within its domain of applicability in this sense, that the interactions and the equations of motion are believed to be correctly described with high approximation. But except for the very lightest atoms the many-particle equations for the constituent electrons are too complicated to be solved otherwise than by relatively crude approximations. The guiding principle of such an approximation is borrowed from, or if one likes forms the basis of, an atomic model, e.g. the Thomas-Fermi model or the Hartree-Fock model. Similarly, various models are used for chemical binding, the metallic state, etc.

Except for the very lightest nuclei, the atomic nucleus also forms a many-particle problem. Here the interactions between the constituent parts, viz. the nuclear forces between the nucleons (protons and neutrons), are only roughly known. In fact the picture of the nucleus built up from interacting nucleons is itself a simplified model. So is the picture of an atom built up from interacting electrons and a nucleus. But the underlying fundamental theory of quantum electrodynamics is in a less unsatisfactory state than the underlying meson theory in the case of nuclear structure. As an intermediate stage between the incomplete theory of interacting nucleons and the observed phenomena of the nuclei there is a variety of nuclear models, e.g. the liquid drop model, the compound nucleus model, the shell model, the collective model, the optical model, the many-particle model of nuclear matter. In some aspects various models are growing towards each other, in other aspects they appear complementary. They serve as an intermediate stage in both directions. On the one hand they are used to deduce properties of the nuclei from the theoretical concepts, on the other hand they provide possibilities of obtaining more information about nuclear forces and nuclear structure from the observed properties. One is not always sure whether the difficulties one encounters are not introduced by the approximations of the model.

These examples show that there is no sharp distinction between a substitute model and a fundamental theory. Every fundamental theory is in its turn an approximation and may appear as a substitute model in a later stage of the theoretical development. Also the distinction between the more abstract character of the fundamental theory and the more pictorial character of the substitute model is relative and may shift with the adaptation of physicists to the evolution of theoretical ideas.

III. STUDY MODEL

Most of the models discussed so far intend to give what might be called in a philosophically rather uncritical way a more or less approximate representation of physical reality. There are other models which deliberately omit some essential realistic aspects in order to study some other aspects independently. Various models of this kind have been studied, e.g. in quantum field theory. Difficulties with divergencies, renormalization, ghost states, etc., are then studied in formalisms like the Lee model and modifications, which are considerably simpler than those of realistic fields. They are intended to provide better insight and more experience for the still more difficult realistic problems.

Study models may also be used for a critical examination of fundamenta concepts. Just as a representative model may give much greater flexibility for changing experimental conditions, so may a study model give possibilities for changing theoretical conditions even in a perhaps unrealistic way. Let us very schematically consider the electromagnetic field. The instantaneous Coulomb interaction at a distance at the beginning of electricity appeared unsatisfactory. Several mechanical models were tried to bridge the gap. Then the ether was introduced as a model of a continuous medium for the electromagnetic field. Under non-stationary conditions the field had a finite velocity of propagation. The ether concept provided for the pictorial feature of the model, but the field became more and more abstract. In a similar way as the concept of 'absolute upright' has lost its meaning on the spherical earth, the concept of 'absolute rest' lost its meaning in special relativity theory and so did the ether concept. Therewith the model lost its pictorial base. For some people this is still indigestible, as can be seen in discussions about the clock paradox. Now under very special conditions the electromagnetic field may be eliminated from the theory and then one is left with an interaction between electric charges at different places and different times. Then this interaction may be changed in various ways so that it is no longer possible to reintroduce a field by which this interaction may be represented. In this way one obtains a study model which on the one hand clearly enlightens some important functions of the field concept in the usual theory and on the other hand shows some restrictions which the use of the field concept imposes upon possible modifications of the theory.

It need hardly be repeated that there is not always a sharp distinction between a study model and other types of models or fundamental theory.

IV. PICTURE, MODEL AND THEORY

If we think of Atlas carrying the sky, it is quite arbitrary, though perhaps useful, to make a further distinction and to speak of a picture rather than a model. As compared with a model a picture is often more easily accepted on a naive level, and in spite of realistic intentions it may have symbolic features. One may observe in the development of physics a gradual shift from pictures to models and from models to theories. At the same time there is a shift from a more pictorial towards a more abstract character. This requires more abstract and formal means of expression and communication. Still there remains an important function for everyday language and for common pictures, although they are no longer adequate in all respects.

This development is clearly shown by quantum mechanics. Compared with classical mechanics the theoretical formalism is more abstract and its relation to observation is less direct. In classical theory one has particles and waves and these two concepts mutually exclude each other. In quantum theory the same entities, e.g. electrons or photons, show particle aspects and wave aspects in the peculiar sense of complementarity – they exclude each other and complete each other. Even today this is considered in some quarters as an inconsistency of quantum theory. In fact the formalism in this sense is entirely consistent and complete. But for greater ease of expression and communication one may speak in terms of particles and of waves as pictures with which we are familiar from daily experience and from classical theory. The formalism may be expressed in various equivalent representations. In treating a special problem it is often useful to transform from one particular representation to another one at appropriate moments. Now the particle picture and the wave picture are inadequate approximations to the formalism. Being different approximations to different representations they exclude each other and complete each other. In speaking in terms of them during the treatment of a special problem it is often necessary to jump from one picture to another one at appropriate moments.

The formal theoretical description of a physical process during the course of time may in classical theory be continuously accompanied by a picture corresponding to continuous observation of the process in terms of the classical concepts of daily experience. This is no longer possible in quantum theory. Here the rules of correspondence between observations and the formalism are such that the formalism is switched on after an observation and switched off before the next one. In between are logical operations in the formalism which have no counterpart in observational pictures. If another observation were to be inserted between the two, the logical operations in the formalism and the (statistical) relations between the two observations would be essentially changed. The continuous one-to-one correspondence between the abstract formalism and the observational picture is lost in quantum theory. Some physicists try to find a quantum theory in which such a correspondence is regained. Up to now no one has succeeded and I do not expect that anyone will.

In a different sense than that of pictures of individual particles and waves, the concepts of various kinds of elementary particles like electrons and photons and of various kinds of basic fields like the electron field and the photon field still have a fundamental meaning in the more elementary parts of quantum theory. In more advanced parts they appear as substitute models. Here the fundamental theory is still in the making.

V. EXPLANATORY MODEL

It might be objected that I have failed to appreciate the function of the model in physics not merely as a substitute for a fundamental theory, but as a genuine explanation. It may be the case that many models have been intended to provide such an explanation, just as this may be the case for many fundamental theories. Apart from some models already discussed, one might mention mechanical models of sound or heat or in general of statistical mechanics. But I think that if there is any genuine explanatory function, it has gradually been shifted from the model to the fundamental theory. In practice the explanatory function of models appears rather obsolete in present-day physics. I am happy that a pseudo-discussion of the explanatory function of fundamental theories may be considered to lie beyond my present topic.

University of Groningen

AXIOMATIZABILITY OF GEOMETRY WITHOUT POINTS

The aim of this paper is to make more precise the well-known conviction that geometry may be built without speaking about points. In the first section we prepare some general syntactical theorems which are needed. In the second section we apply these theorems to a certain theory of topology without points.

I. SOME THEOREMS ON AXIOMATIZABILITY AND INTERPRETATION

We shall consider theories formalized in the lower predicate calculus. Every theory will be considered as containing only theorems without free variables. By a sentence I mean only a closed one. Cn(X) will denote the set of all logical consequences of the set X; X and Cn(X) are sets of sentences.

Let $\mathrm{Cn}_{R_1R_2}(X)$ be the set of all logical consequences of the set X containing only the predicates R_1, R_2 and with all quantifiers relativized to the predicate R_1. (After a general quantifier $\bigwedge x$ there is always an implication $(R_1(x) \rightarrow \ldots)$ and after an existential quantifier $\bigvee x$ stands always a conjunction $(R_1(x) \wedge \ldots)$.

A set X is a theory with standard relativized formalization with respect to the predicates R_1, R_2 if and only if

(1) $$X = \mathrm{Cn}_{R_1R_2}(X).$$

Let D_R^A be the definition of the predicate R by means of the definiens A:

(2) $$D_R^A = \ulcorner \bigwedge x, \ldots y\, (R(x, \ldots y) \equiv A) \urcorner.$$

We shall say that a theory $X_{R_1R_2}$ has an interpretation in $Y_{K_1K_2}$ by means of the definiens A_1, A_2, if and only if A_1 and A_2 contain only the predicates K_1 and K_2, all quantifiers in A_1, A_2 are relativized to K_1 and:

(3) $$X_{R_1R_2} \subset \mathrm{Cn}_{R_1R_2}(Y_{K_1K_2} \cup \{D_{R_1}^{A_1}, D_{R_2}^{A_2}\}).$$

Let is $(R_1, R_2, R'_1, R'_2, F)$ be the following formula:

$$
(4) \qquad \begin{aligned}
&\bigwedge x \left(R_1(x) \rightarrow \bigvee y(R'_1(y) \wedge F(x, y))\right) \wedge \\
&\bigwedge y \left(R'_1(y) \rightarrow \bigvee x(R_1(x) \wedge F(x, y))\right) \wedge \\
&\bigwedge x, y, z \left((F(x, y) \wedge F(x, z)) \rightarrow y = z\right) \wedge \\
&\bigwedge x, y, z \left((F(x, y) \wedge F(z, y)) \rightarrow x = z\right) \wedge \\
&\bigwedge x, y, z, u \left((F(x, y) \wedge F(z, u)) \rightarrow (R_2(x, z) \equiv R'_2(y, u))\right)
\end{aligned}
$$

where F is a formula with two free variables.

We shall consider some theories with standard relativized formalizations. Let $T_{P_1P_2}$, $S_{Q_1Q_2}$, $T'_{P'_1P'_2}$, $S'_{Q'_1Q'_2}$ be such theories with respect to the predicates respectively P_1, P_2; Q_1, Q_2,; P'_1, P'_2; Q'_1, Q'_2. P_1, Q_1, P'_1, Q'_1 are one variable predicates, P_2, Q_2, P'_2, Q'_2 are two-variable predicates. In order to simplify the formulas we shall assume that the letters T, S, T', S' denote respectively the theories $T_{P_1P_2}$, $S_{Q_1Q_2}$, $T'_{P'_1P'_2}$, $S'_{Q'_1Q'_2}$. T' is the analogue of T and S' is the analogue of S (i.e. there is a one-one mapping of T on T' and of S on S' consisting of adding the sign ′ to the predicates of T (or of S)).

Lemma 1. If E is a sentence containing only the predicates Q_1, Q_2, all quantifiers in E are relativized to Q_1 and E' is the analogue of E, then for every F,

$$
(5) \qquad \ulcorner E \equiv E' \urcorner \in \mathrm{Cn}\,\{\mathrm{is}\,(Q_1, Q_2, Q'_1, Q'_2, F')\}
$$

Proof. By induction. Lemma 1 is a syntactical formulation of a familiar property of isomorphism.

Theorem I. Under the above assumptions concerning the theories T, S, T', S', if S has an interpretation in T by A_1 and A_2, T' has an interpretation in S by B_1 and B_2, and for some formula F with two free variables:

$$
(6) \qquad \mathrm{is}\,(Q_1, Q_2, Q'_1, Q'_2, F) \in \mathrm{Cn}\left(S \cup \{D^{B_1}_{P'_1}, D^{B_2}_{P'_2}, D^{A'_1}_{Q'_1}, D^{A'_2}_{Q'_2}\}\right),
$$

then $\mathrm{Cn}_{Q_1Q_2}\left(T \cup \{D^{A_1}_{Q_1}, D^{A_2}_{Q_2}\}\right) \subset S$.

Proof. If $E \in \mathrm{Cn}_{Q_1Q_2}(T \cup \{D^{A_1}_{Q_1}, D^{A_2}_{Q_2}\})$, then by analogy:

$$
(7) \qquad E' \in \mathrm{Cn}_{Q'_1Q'_2}\left(T' \cup \{D^{A'_1}_{Q'_1}, D^{A'_2}_{Q'_2}\}\right).
$$

On the other hand if T' has an interpretation in S by B_1 and B_2, then by (3):

$$
(8) \qquad T' \subset \mathrm{Cn}_{P'_1P'_2}\left(S \cup \{D^{B_1}_{P'_1}, D^{B_2}_{P'_2}\}\right).
$$

From (7) and (8) it follows that

(9) $$E' \in \mathrm{Cn}_{Q'_1 Q'_2}\left(S \cup \{D^{B_1}_{P'_1}, D^{B_2}_{P'_2}, D^{A'_1}_{Q'_1}, D^{A'_2}_{Q'_2}\}\right).$$

(5), (6) and (9) imply that

(10) $$E \in \mathrm{Cn}_{Q_1 Q_2}\left(S \cup \{D^{B_1}_{P'_1}, D^{B_2}_{P'_2}, D^{A'_1}_{Q'_1}, D^{A'_2}_{Q'_2}\}\right).$$

But E does not contain the predicates P'_1, P'_2, Q'_1, Q'_2. Hence (by the theorem on the elimination of definitions) (10) implies that $E \in \mathrm{Cn}_{Q_1 Q_2}(S) = S$, which completes the proof.

Lemma 2. If A_1 and A_2 contain only the predicates P_1, P_2 and all quantifiers in A_1 and A_2 are relativized to P_1, and if A'_1, A'_2 are the analogues of A_1, A_2, then for every formula F there exists a formula G with two variables such that

$$\mathrm{is}\,(Q_1, Q_2, Q'_1, Q'_2, G) \in \mathrm{Cn}\,(\mathrm{is}\,(P_1, P_2, P'_1, P'_2, F), D^{A_1}_{Q_1}, D^{A_2}_{Q_2}, D^{A'_1}_{Q'_1}, D^{A'_2}_{Q'_2})$$

Proof. By induction with respect to the shape of the formulas A_1 and A_2. Lemma 2, like lemma 1, may be considered as an exact formulation of a certain well-known property of isomorphism.

Theorem II. If T is finitely axiomatizable, S has an interpretation in T by A_1 and A_2, T' has an interpretation in S by B_1 and B_2 and for some formula F with two free variables

(11) $$\mathrm{is}\ \ (P_1, P_2, P'_1, P'_2, F) \in \mathrm{Cn}\left(T \cup \{ D^{A_1}_{Q_1}, D^{A_2}_{Q_2}, D^{B_1}_{P'_1}, D^{B_2}_{P'_2}\}\right)$$

then the set $\mathrm{Cn}_{Q_1 Q_2}\left(T \cup \{ D^{A_1}_{Q_1}, D^{A_2}_{Q_2} \}\right)$ is finitely axiomatizable.

Moreover we can describe effectively the axioms of the theory $\mathrm{Cn}_{Q_1 Q_2}(T \cup \{ D^{A_1}_{Q_1}, D^{A_2}_{Q_2} \})$. It suffices to describe the axiom system using the definitions. Such an axiom system may consist of:

(i) The analogues Ax' of the axioms Ax of T,

(ii) the sentence is $(Q_1; Q_2, Q'_1, Q'_2, G)$ for some G,

(supposing the definitions: $D^{A_1}_{Q_1}, D^{A_2}_{Q_2}, D^{B_1}_{P'_1}, D^{B_2}_{P'_2}, D^{A'_1}_{Q'_1}, D^{A'_2}_{Q'_2}$).

Proof. Ax' is the consequence of T using the definitions because $Ax' \in T'$ and T' has the interpretations in T by B_1 and B_2. is $(Q_1, Q_2, Q'_1, Q'_2, G)$ is the consequence of T and the definitions according to (11) and lemma 2. Now let X be a finite set of sentences equivalent to (i) $\cup$ (ii) (using the six definitions mentioned above) and containing only the predicates Q_1 and Q_2 and with all quantifiers relativized to Q_1. We shall consider the set $\mathrm{Cn}_{Q_1 Q_2}(X)$. From the above definition of the set X and from the theorem on the elimination of the definitions it follows that

$$\mathrm{Cn}_{Q_1Q_2}(X) \subset \mathrm{Cn}_{Q_1Q_2}\left(T \cup \{D^{A_1}_{Q_1}, D^{A_2}_{Q_2}\}\right).$$

This means that $\mathrm{Cn}_{Q_1Q_2}(X)$ has an interpretation in T. On the other hand (i) implies that T' has an interpretation in $\mathrm{Cn}_{Q_1Q_2}(X)$ by B_1 and B_2. From (ii) it follows that is $(Q_1, Q_2, Q'_1, Q'_2, G) \in \mathrm{Cn}\,(\mathrm{Cn}_{Q_1Q_2}(X) \cup \{D^{B_1}_{P'_1}, D^{B_2}_{P'_2}, D^{A'_1}_{Q'_1}, D^{A'_2}_{Q'_2}\})$.

Hence all conditions of Theorem I are satisfied, and according to Theorem I

$$\mathrm{Cn}_{Q_1Q_2}(X) = \mathrm{Cn}_{Q_1Q_2}\left(T \cup \{D^{A_1}_{Q_1}, D^{A_2}_{Q_2}\}\right)$$

which completes the proof.

Both the above theorems may be extended. Instead of the predicate calculus we can take set theory or a logical calculus of higher type with the axioms of identity and extensionality, which are necessary to extend the isomorphism from the lowest type to all higher types (lemmas 1, 2). As logical consequences we shall consider the consequences on the basis of this logical or set-theoretical calculus. Moreover we must change the meaning of the relativization of the quantifiers. Instead of a simple relativization of a variable x to the predicate $R_1(x)$ we must introduce different relativizations for different types, saying that: '$R_1(x)$' or 'the elements of x are R_1, or 'the elements of the elements of x are R_1' etc.

Theorem III. With the above modification Theorems I and II remain true.

It is also obvious that the theorems may be extended to theories containing more than two primitive predicates.

II. A SYSTEM OF TOPOLOGY WITHOUT POINTS

In this section we shall apply the theorems of section 1 to prove the finite axiomatizability of the topology obtained in the pointless geometry. Let T and S be two theories based on the set theory.

The specific axioms of theory T say that:

(i) the basic elements which are named spatial bodies ($P_1(x)$) constitute a mereological field (a boolean algebra without zero-element) [1]) with a primitive predicate $x \subset y$ (the body x is contained in the body y);

(ii) $x)(y$ is the relation of being separated:

[1]) For mereological fields see: H. S. Leonard and N. Goodeman, The Calculus of Individuals and its Uses. Jour. of Symb. Logic 5 (1940), 45–55 and A. Grzegorczyk, The systems of Leśniewski in relation to contemporary logical research. Studia Logica 3 (1955), 77–97.

A1 $x)(y \rightarrow \sim (x \subset y)$
A2 $x)(y \rightarrow y)(x$
A3 $(x)(y \wedge z \subset x) \rightarrow z)(y$
A4 $\big(P_1(x) \wedge P_1(y) \wedge \sim (x)(y)\big) \rightarrow \bigvee a \{\bigwedge x(x \in a \rightarrow$
$P_1(x)) \wedge \bigwedge u, v((u \in a \wedge v \in a) \rightarrow (u = v \vee u M v \vee v M u)) \wedge$
$\bigwedge u, v \big(\bigwedge z(z \in a \rightarrow (z \triangle u \wedge z \triangle v)) \rightarrow \sim (u)(v)\big) \wedge$
$\bigwedge z(z \in a \rightarrow (z \triangle x \wedge z \triangle y)) \wedge \bigvee z(z \in a \wedge (x \subset y \rightarrow z \subset x)) \wedge$
$\bigwedge z(z \in a \rightarrow \bigvee y(y \in a \wedge y M z))\}$
where $x \triangle y$ is the abbreviation of $\bigvee z(z \subset x \wedge z \subset y)$,
and $x M y$ is the abbreviation of $x \subset y \wedge \bigwedge z(\sim (z \triangle y) \rightarrow z)(x)$.

The specific axioms of theory S say that:
(x) the basic elements which are named points ($Q_1(x)$) constitute a topological space with a primitive predicate $Q_2(x)$ (x is an open set of points),
(xx) two distinct points are separable by two disjoint open sets;
(xxx) for every point p there exists an infinite strictly decreasing (M) family of open sets such that the intersection of the family is equal to p.

In T we introduce the following definitions:
$Q(a) \equiv \bigwedge x(x \in a \rightarrow P_1(x)) \wedge \bigwedge u, v\big((u \in a \wedge v \in a) \rightarrow$
$(u = v \vee u M v \vee v M u)\big) \wedge \bigwedge u, v \{\bigwedge z\big(z \in a \rightarrow (z \triangle u \wedge$
$z \triangle v)\big) \rightarrow \sim (u)(v)\} \wedge \bigvee z\, z \in a \wedge$
$\bigwedge z(z \in a \rightarrow \bigvee x\,(x \in a \wedge x M z)\big).$
D_1 $Q_1(a) \equiv \bigvee b \{ Q(b) \wedge \bigwedge x \big(x \in a \equiv (P_1(x) \wedge \bigvee y(y \in b \wedge y \subset x)\big)\}.$
D_2 $Q_2(a) \equiv \bigwedge x(x \in a \rightarrow Q_1(x)) \wedge \bigwedge b(b \in a \rightarrow \bigvee x(x \in b \wedge$
$\bigwedge c((Q_1(c) \wedge x \in c) \rightarrow c \in a)$

In S we introduce the following definitions:
$p \in \text{Int}\,(x) \equiv Q_1(p) \wedge \bigwedge a(a \in x \rightarrow Q_1(a)) \wedge \bigvee y(Q_2(y) \wedge y \subseteq x \wedge p \in y)$
$C(x) = Q_1 - \text{Int}\,(Q_1 - x)$
where $\subseteq$ denotes set-theoretical inclusion and — denotes set-theoretical complementation.
D_3 $P'_1(x) \equiv Q_2(x) \wedge x = \text{Int}\,(C(x)) \wedge \bigvee p\, p \in x$
D_4 $x \subset' y \equiv P'_1(x) \wedge P'_1(y) \wedge x \subseteq y$
D_5 $x)('y \equiv P'_1(x) \wedge P'_1(y) \wedge \bigwedge p(p \in C(x) \rightarrow p \notin C(y))$

Theorem IV. Theory S contains all set-theoretical consequences of theory T containing the predicates Q_1 and Q_2.

Proof. In order to apply Theorem 2 we must point out some consequences of the axioms of T.

Using only the definitions of Q_1 and Q_2 it is easy to prove that the set of all points and the empty set are open ones. The elementary boolean properties of $\subset$ imply that the intersection of two open sets is an open set, and the union of a family of open sets is an open set. Next we define the set of internal points of a body:

(1) $$p \in \mathrm{Irl}(x) \equiv (P_1(x) \wedge Q_1(p) \wedge x \in p).$$

The set of internal points is an open one:

(2) $$P_1(x) \rightarrow Q_2(\mathrm{Irl}(x)).$$

This follows from the definitions and the elementary properties of $\subset$. From axioms A1 and A4, setting $y = x$, we obtain the result that the set of internal points is non-empty:

(3) $$P_1(x) \rightarrow \bigvee p(p \in \mathrm{Irl}(x)).$$

The sets $\mathrm{Irl}(x)$ constitute the family of environments. If $p \neq q$ are two distinct points then there exists a body $x \in p$ such that $x \notin q$ (or conversely). We shall prove that there exist two bodies u, z such that $u \in p$ and $z \in q$ and $\sim (u \triangle z)$. Let us suppose that for every body $z \in q$, $z \triangle x$. The set q is non empty, then for some $z_0 \in q$ we have that for every $z \in q$, $z \triangle x$ and $z \triangle (z_0 - x)$. When for some $z \in q$ would be $z \subset x$, x would belong to q, then $z_0 - x$ is a body and for every $z \in q$, $z \triangle (z_0 - x)$. The definition of Q_1 implies that there exists a body $u \in p$ such that $u \, M \, x$. Hence from the definition of M it follows that $u)(z_0 - x$, because $\sim (z_0 - x \triangle x)$. Thus the definitions of Q_1 and Q imply that there exists $z \in q$ such that $\sim (z \triangle u)$. The sets $\mathrm{Irl}(u)$ and $\mathrm{Irl}(z)$ are two disjoint open environments which separate the points p and q. Then the topological axioms (x) and (xx) of S are obtained in T.

Now we need the following characterization of the closure:

(4) $$p \in C\,X \equiv Q_1(p) \wedge \bigwedge x(x \in p \rightarrow \bigvee q(q \in X \wedge x \in q))$$

The first implication ($\rightarrow$) follows from the fact that the set $\mathrm{Irl}(x)$ for $x \in p$ is an open environment of p; hence there exists a point $q \in x$ such that $q \in \mathrm{Irl}(x)$, what means that $x \in q$. Conversely for every Y, if $p \in Y$ and Y is an open set, then according to the definition of Q_2 there exists a body $x \in p$ such that $\mathrm{Irl}(x) \subseteq Y$. Hence if p satisfies the right-hand part of the above equivalence, then there exists a point $q \in X$ such that $x \in q$, what means that $q \in \mathrm{Irl}(x)$, and thus $q \in Y$. Hence $p \in X$, by the familiar properties of closure.

The formulas (3) and (4) imply the following one:

$$p \in C\,\mathrm{Irl}(y) \equiv Q_1(p) \wedge \bigwedge x(x \in p \rightarrow x \bigtriangleup y) \tag{5}$$

From the axiom A4, (3) and (5) we obtain the following characterization of two primitive notions of the theory T:

$$x \subset y \equiv \mathrm{Irl}(x) \subseteq \mathrm{Irl}(y) \tag{6}$$

$$x)(y \equiv C\,\mathrm{Irl}(x) \cup C\,\mathrm{Irl}(y) = 0 \tag{7}$$

and a similar characterization of M:

$$x\,M\,y \equiv C\,\mathrm{Irl}(x) \subseteq \mathrm{Irl}(y) \tag{8}$$

The definition of Q_1, (8) and (xx) imply (xxx). Indeed if $Q_1(p)$, $Q(a)$ and $\bigwedge x(x \in p \equiv \bigvee y(y \in a \wedge y \subset x))$, then the family of open sets $\mathrm{Irl}(x)$ for $x \in a$ constitute a strictly decreasing (M) family of open sets with intersection equal to p. Hence the theory S has the interpretation in T by means of the definitions D_1 and D_2.

On the other hand let T' be the theory obtained from T by adding $'$ to the primitive terms. It is easy to verify that T' has the interpretation in S by means of the definitions D_3 — D_5. The open non-empty domains P'_1 constitute a mereological field with the primitive relation $\subset'$.[1] The axioms A1 — A4 are also evidently satisfied for $)('$ defined in D_5.

Now in order to apply the Theorems of the preceding section we need to prove in T the theorem establishing the isomorphism between the bodies and the open domains of their internal points. Indeed the set of internal points is an open domain:

$$\mathrm{Int}(C(\mathrm{Irl}(x))) \subseteq \mathrm{Irl}(x). \tag{9}$$

Namely if $p \in \mathrm{Int}(C(\mathrm{Irl}(x)))$, then by the definition of Int there exists an open set Y such that $p \in Y \subseteq C(\mathrm{Irl}(x))$; hence by D_2 there exists a body y such that $p \in \mathrm{Irl}(y)$ and

[1]) For open and closed domains see: K. Kuratowski, Topologie I (Warszawa 1956), 42, 43.

(10) $$\mathrm{Irl}(y) \subseteq C(\mathrm{Irl}(x)).$$

The formulas (10) and (5) imply that:

(11) for every body z, if $z \subset y$ then $z \triangle x$.

Hence from the general boolean theorems, $y \subset x$ and thus $p \in \mathrm{Irl}(x)$.
The inclusion converse to (9) is generally true for every open set, and so also for $\mathrm{Irl}(x)$ according to (2).
From D_3, (2), (3) and (9) we obtain the result that:

(12) $$P_1(x) \rightarrow P'_1(\mathrm{Irl}(x)).$$

On the other hand we can prove in T that:

(13) $$P'_1(X) \rightarrow \bigvee x(P_1(x) \wedge X = \mathrm{Irl}(x))$$

According to D_3 if $P'_1(X)$ then there exists a point p such that $p \in X$. X is an open set, then according to the definition D_2 there exists a body z such that: $p \in \mathrm{Irl}(z)$ and $\mathrm{Irl}(z) \subseteq X$. Hence there exists a body x which is the boolean union of all bodies z such that $\mathrm{Irl}(z) \subseteq X$. It is easy to prove that $X = \mathrm{Irl}(x)$. Thus according to (6), (7), (12) and (13), the relation F:

$$F(x, y) \equiv \big(P_1(x) \wedge P'_1(y) \wedge y = \mathrm{Irl}(x)\big)$$

establishes an isomorphism between the bodies $P_1($ and the open non empty domains $P'_1)$.
Now all conditions of Theorem II are satisfied; thus the set of topological consequences of T is finite axiomatizable and it is easy to prove that the axioms pointed by Theorem II are consequences of the axioms (x) — (xxx) of S.
It is possible also to interpret bodies as closed non-empty domains. But if we wish bodies to satisfy the axioms of boolean algebra or the axioms of mereology then these are the only two possibilities.

University of Warsaw

MARCEL GUILLAUME

CERTAINS ASPECTS SYNTACTIQUES D'UNE NOTION DE MODÈLE: RELATIVISATION D'UNE FONCTION LOGIQUE DE CHOIX [1])

Ce que je veux dire concerne un mode selon lequel la notion de modèle apparaît en mathématique depuis fort longtemps, avant qu'il ait été généralisé et codifié selon des règles syntactiques précises par différents auteurs et sous différents noms. On peut voir en effet dans le procédé dont je vais parler une variante du procédé de *translation* d'une théorie dans une autre de Wang (7); cependant, comme j'en délaisserai les cas les plus complexes pour en éviter les complications, souvent inessentielles, on pourra aussi, en regardant bien, en rattacher les cas que j'envisagerai au procédé d'*interprétation* de Tarski (6). Enfin, on verra aussi comment ce procédé s'apparente étroitement au procédé de *relativisation* employé par Gödel dans son ouvrage célèbre sur la compatibilité de l'axiome du choix (4).

C'est d'ailleurs toujours à l'appui d'une démonstration de non-contradiction relative, à la fin du compte, que la notion de modèle intervient en mathématique selon ce mode, ici un peu rafraîchi par son adaptation au système entre théories duquel ceci prépare des démonstrations de non-contradiction relative: le système formalisé de Bourbaki, à peu de choses près tel qu'on le trouve dans le Chapitre I du Livre de Théorie des Ensembles de cet auteur (2), avec, en particulier, sa fonction logique de choix de Hilbert, son utilisation de celle-ci pour la définition des quantificateurs, et son absence de quantificateurs primitifs. En principe, j'adopterai la terminologie et les notations de cet ouvrage, quels que puissent en être les défauts; la discussion de ce point nous ferait sortir du cadre de nos préoccupations présentes.

Je supposerai, cependant, que les théories dont il sera question sont toutes dénuées de signes substantifiques autres que leurs *constantes*, qui sont des signes substantifiques de poids 0; on peut toujours se ramener à ce cas, par un procédé bien connu, qui prend en l'occurrence une forme très simple; cette hypothèse amène des simplifications appré-

[1]) Les chiffres entre parenthèses renvoient aux items de la Bibliographie.

ciables. Il est sans doute utile de rappeler que Bourbaki utilise pour désigner sa fonction logique de choix le symbole τ employé par Hilbert dans ses premiers écrits sur ce sujet (3), et qu'il réserve le nom de 'constantes' d'une théorie aux lettres qui figurent dans ceux des axiomes de cette théorie qui ne sont pas fournis par un schéma; il sera commode d'appeler les autres lettres des 'variables' de cette théorie, d'appeler les assemblages figurant dans au moins une construction formative des *assemblages bien formés*, puis *relations bien formées* ce que Bourbaki nomme 'relations', et qui est connu de tout le monde sous le nom de 'Wffs'; et enfin, d'appeler *relations closes* et *termes clos*, respectivement, les relations (bien formées) et les termes où ne figure *aucune variable*.
Ayant en vue de dégager progressivement, pour des théories de cette espèce, des instructions syntactiques rendant compte de raisonnements tenus en fait sur des modèles à laisser en définitive presqu'entièrement sous-entendus, j'utiliserai une notion intuitive de modèle pour ce genre de théories, sans chercher à donner à son sujet de précisions de nature sémantiques, qui seraient analogues à celles qu'on peut trouver, par exemple, dans une étude de Günter Asser (1).

Dans ce qui suit maintenant, on dispose d'un modèle M d'une certaine théorie quantifiée $\mathfrak{T}$, et on se propose d'utiliser ces données pour 'bâtir' un modèle $M°$ d'une autre théorie quantifiée $\mathfrak{T}°$, en prenant pour domaine d'objets de $M°$, un domaine partiel du domaine d'objets de M, et pour relations liant ces objets, certaines des relations liant les objets de M, comprises comme opérant dans le domaine partiel auquel on porte attention. Toute notion de $\mathfrak{T}°$ dans son modèle $M°$ *sera* donc une notion de $\mathfrak{T}$ dans son modèle M; comme toute notion de $\mathfrak{T}°$, qui fait dans $\mathfrak{T}°$ l'objet d'une définition, est déterminée, en dernière analyse, par les notions primitives de $\mathfrak{T}°$, la détermination des notions primitives de $\mathfrak{T}°$ dans le modèle $M°$, au moyen des notions de $\mathfrak{T}$ dans le modèle M, entraîne celle de toutes les notions de $\mathfrak{T}°$ dans le modèle $M°$, au moyen de notions de $\mathfrak{T}$ dans le modèle M.
Si nous voulons rester sur le plan formel, syntactique, les modèles restent sous-entendus, et donc, disparaissent, entraînant avec eux les notions proprement dites dont il ne reste que les assemblages de signes qui les représentent. De la condition selon laquelle une notion de $\mathfrak{T}°$ dans le modèle $M°$ est une notion de $\mathfrak{T}$ dans le modèle M, il ne pourra

subsister que la condition selon laquelle un assemblage a de $\mathfrak{T}^\circ$ représente la même chose que certain assemblage a° de $\mathfrak{T}$. Intuitivement, ce résidu syntactique pourra encore s'exprimer en disant qu'on *interprète* les assemblages de $\mathfrak{T}^\circ$ comme étant des assemblages convenables de $\mathfrak{T}$; il sera commode de continuer à appeler l'assemblage a° *l'interprétation* de l'assemblage a.

De la condition selon laquelle les objets de M° sont certains des objets de M, il ne pourra pareillement subsister que la condition selon laquelle tout terme de $\mathfrak{T}^\circ$ a pour interprétation un terme de $\mathfrak{T}$, répondant aux conditions imposées pour désigner un objet de M°. De même, le résidu syntactique de la condition selon laquelle les relations liant les objets de M° sont certaines des relations liant les objets de M sera la condition selon laquelle toute relation bien formée de $\mathfrak{T}^\circ$ a pour interprétation une relation bien formée de $\mathfrak{T}$.

On dispose pour réaliser ce programme d'une très grosse marge de liberté, même si, comme dans la suite, on s'en tient à un 'principe de permanence' selon lequel la détermination, au moyen de notions de $\mathfrak{T}$ dans le modèle M, d'une notion primitive de $\mathfrak{T}^\circ$ dans le modèle M°, *ne doit pas dépendre du contexte* dans lequel elle figure. Par exemple, une relation bien formée devant représenter une relation d'appartenance devra reçevoir, conformément à ce principe, une définition en toute généralité, indépendante des objets que peuvent représenter les termes y figurant dans les rôles respectifs de l'élément et de l'ensemble. Le résidu syntactique de ce principe consiste à poser que l'interprétation de tout assemblage bien formé $\sigma a_1 \ldots a_n$, où les a_i sont des assemblages bien formés et σ un signe de poids n (le poids 0 étant affecté aux lettres de toutes natures, le poids 1 aux signes $\neg$ et τ, et le poids 2 à v), est l'assemblage bien formé $\beta_1 \mathring{a}_{i_1} \beta_2 \mathring{a}_{i_2} \ldots \beta_m \mathring{a}_{i_m} \beta_{m+1}$, où tout $\mathring{a}_{i_j}$ est l'interprétation de l'assemblage a_{i_j}, qui fait partie des a_i, et où, ne dépendant des a_i ni par leur nombre, ni par leur ordre, ni par leur composition ou par leur valeur, les assemblage β_j (dont certains peuvent être vides, mais pas tous) et les i_j constituent ce qu'il sera commode d'appeler aussi l'*interprétation* σ° de σ.

Ainsi, une partie de ce qui subsiste sur le plan formel de l'élaboration du modèle M° prend l'aspect d'un *procédé de transposition* d'assemblages de $\mathfrak{T}^\circ$ en assemblages de $\mathfrak{T}$ faisant des *signes* de $\mathfrak{T}^\circ$ des *symboles abréviateurs* relatifs à la théorie $\mathfrak{T}$.

En ce qui concerne les notions de *négation* et de *disjonction*, tenons-nous en désormais à la plus simple des déterminations répondant aux conditions énoncées: celle qui les fait, prises en tant que notions de $M°$, se déterminer par elles-mêmes, prises en tant que notions de M. Ce qu'il en reste sur le plan syntactique consiste à poser que $\neg°$ n'est autre que $\neg$ et que $v°$ n'est autre que v. Il en résulte aussitôt, en vertu du principe de permanence, que *l'interprétation d'un exemple quelconque d'un schéma de fonction propositionnelle de $\mathfrak{T}°$ est encore un exemple de ce schéma*, pris en tant que schéma de fonction propositionnelle de $\mathfrak{T}$; les interprétations des exemples des schémas de Hilbert et Ackermann, *S1*, *S2*, *S3* et *S4*, sont donc des *théorèmes* (même des *axiomes*) de $\mathfrak{T}$.
En ce qui concerne les *variables*, acceptons désormais le principe selon lequel une variable doit indiques, dans la théorie $\mathfrak{T}°$ du modèle $M°$, une indétermination *du même ordre* que dans la théorie $\mathfrak{T}$ du modèle M, bien que la *marge* de cette indétermination doive être *illimitée* dans $M°$, et *limitée*, dans M, au domaine partiel auquel on porte attention (Cette différence de marge échappe aux procédés d'interprétation mis en oeuvre, puisqu'elle est du ressort des jugements de valeur que nous portons sur la façon dont les relations lient les objets des modèles; on peut toujours définir, autre chose est de fixer le domaine des objets au sein duquel on se bornera à apprécier l'effet des définitions). Comme il est *loisible* d'admettre que les variables de $\mathfrak{T}°$ et celles de $\mathfrak{T}$ sont les mêmes, la façon la plus simple de respecter ce principe et les précédents consiste à prendre toute varaible x de $\mathfrak{T}$ pour interprétation $x°$ de x elle-même, prise en tant que variable de $\mathfrak{T}°$; je m'en tiendrai à cette pratique dans la suite.
Le dernier principe invoqué sera d'ailleurs aussi supposé satisfait en ce qui concerne les *constantes* de $\mathfrak{T}°$; le résidu syntactique en est la condition selon laquelle l'interprétation de toute constante de $\mathfrak{T}°$ devra être un terme de $\mathfrak{T}$ désignant un objet déterminé sans aucune ambiguité, comme il sied à un être constant, c'est-à-dire, un *terme clos* de $\mathfrak{T}$.
De petites difficultés techniques sont liées à l'usage du signe $\square$ en présence du τ; la façon la plus simple de les éviter est d'assimiler le $\square$ à une lettre, qu'on ne rangera cependant ni parmi les variables, ni parmi les constantes, et à le prendre lui-même pour $\square°$.
Ces règles d'interprétation, jointes à celle qui traduit le principe de permanence, entraînent d'ores et déjà que *l'interprétation de tout assemblage bien formé a de $\mathfrak{T}°$ comporte exactement les mêmes variables que a*

lui-même; que *si* T *est un terme,* x *une variable, et* B *un terme* (resp. *une relation bien formée*), *l'interprétation* $((T1\ x)B)°$ *du terme* (resp. *de la relation bien formée*) $(T1x)B$ *est le terme* (resp. *la relation bien formée*) $(T°1x)B°$; en particulier, *l'interprétation* $((\exists x)R)°$ *d'une relation bien formée* $(\exists x)R$ *de* $\mathfrak{T}°$ *n'est autre que la relation bien formée* $((\tau_x(R))°1x)R°$.
L'élaboration des traits généraux de $M°$ en tant que modèle d'une théorie *quantifiée* va désormais s'achever par la détermination de la fonction logique de choix de Hilbert de $\mathfrak{T}°$ dans le modèle $M°$; or, toujours pour les mêmes raisons, *l'interprétation d'un terme commençant par un* τ *ne peut être qu'un terme commençant par un* τ (puisqu'on délaisse le cas où il pourrait y avoir des signes substantifiques) *portant sur un exemple d'un certain schéma de relation attaché au modèle* $M°$, *exemple que l'interprétation de la relation bien formée sur laquelle porte le* τ *initial achève de déterminer.*

Or, l'interprétation du τ se trouve déjà soumise, en fait, à un certain nombre de conditions, dûes à la façon dont ont été fixées, d'entrée de jeu, les grandes lignes de l'élaboration des traits généraux du modèle $M°$, et qu'un retour sur les intentions qui guident cette élaboration va permettre de dégager. *Comprendre comme opérant dans le domaine partiel* auquel on porte attention les relations de M prises comme relations de $M°$, signifie que l'on tient l'une de ces relations pour 'vraie', dès lors qu'elle est 'vraie' *dans ce domaine partiel seulement;* dès lors, donc, que cette relation *devient* vraie, *lorsque tous les objets qu'elle lie sont du domaine partiel.*
Sur le plan syntactique, il ne reste de cette relation que la relation bien formée qui la représente, où figurent les termes qui représentent les objets qu'elle lie. Pour aller plus loin, il nous faut donc une contrepartie syntactique au fait qu'un terme de $\mathfrak{T}$ désigne un objet du domaine partiel; nous ne la trouverons que si l'on peut *caractériser les objets du domaine partiel par une relation bien formée* A *de* $\mathfrak{T}$, *où figure une variable* x *de* $\mathfrak{T}$ *et une seule, et que vérifient, parmi les termes de* $\mathfrak{T}$, *tous ceux qui désignent un objet du domaine partiel, et eux seulement.*
Supposons donc que tel est le cas; la relation bien formée A est nécessairement telle que $(\exists x)A$ est *valide dans le modèle* M, car sinon, ce qu'on ferait ne serait pas le reflet de la construction effective d'un modèle $M°$; c'est la relation bien formée $(T1x)A$ qui *exprime formellement que le*

terme T désinge bien on objet du domaine partiel, c'est-à-dire de $M°$. Mais, nous voulons rester sur le plan syntactique, et n'expliciter rien de ce qui concerne le modèle M, ni des notions sémantiques qui nous permettraient, en particulier, de caractériser la validité d'une relation bien formée dans un modèle. D'une relation bien formée d'une des théories en cause, nous ne pourrons donc savoir à coup sûr qu'elle est valide dans un modèle de cette théorie, que si elle *en est un théorème*, propriété qui ne caractérise certes pas la validité dans ce modèle, on le sait, mais est la seule dont nous disposions pour l'exprimer sur le plan où nous nous plaçons. Donc, ce n'est qu'en prenant pour A une relation bien formée telle que $(\exists x)A$ soit un *théorème de* $\mathfrak{T}$ que notre façon d'opérer reflètera, sans quitter le plan syntactique, la construction effective d'un modèle $M°$.

Lorsque T est un terme *clos*, la vérification de la condition $(T1x)A$ *dépend essentiellement du modèle M*; donc, pour réaliser ce qui a été décidé, il faudra notamment faire en sorte que, pour tout terme clos T de $\mathfrak{T}°$, la relation bien formée $(T°1x)A$ soit *valide dans le modèle M*; nous venons de voir que cela sera fait comme il a été décidé, c'est-à-dire sans quitter le plan syntactique, si nous agissons de telle sorte que, pour *tout* terme *clos* T de $\mathfrak{T}°$, la relation bien formée $(T°1x)A$ soit un *theorème de* $\mathfrak{T}$. Dans ces conditions, tous les termes qui figurent dans une relation close $R°$ de $\mathfrak{T}$ interprétant une relation close R de $\mathfrak{T}°$ désignent nécessairement des objets du domaine partiel, et cette relation *close R* sera tenue pour *valide dans le modèle M°*, si et seulement si son interprétation $R°$ est *valide dans le modèle M*; le résidu syntactique de cette condition sera que nous aurons à tenir pour *valide dans le modèle M°* toute relation *close* dont *l'interprétation* est un *théorème de* $\mathfrak{T}$ – avec le déchet correspondant à l'impossibilité de demander la réciproque, en général.

Par contre, si un terme se réduit à une variable y, la marge d'indétermination dans M représentée dans $\mathfrak{T}$ par son interprétation y est illimitée, et pour être satisfaite, la condition $(y1x)A$ exprimant qu'il désigne un objet du domaine partiel doit en général être *posée*. Le résidu syntactique de cet acte de l'esprit consiste à ajouter l'hypothèse $(y1x)A$ (entre autres, peut-être), c'est-à-dire à passer dans une théorie $\mathfrak{T}''$ plus forte que la théorie $\mathfrak{T}'$ obtenue à partir de $\mathfrak{T}$ par adjonction de l'axiome supplémentaire $(y1x)A$. Mais le passage de $\mathfrak{T}$ à $\mathfrak{T}'$ a pour effet de *transformer y en une constante de* $\mathfrak{T}'$; donc, une relation bien formée et un terme de $\mathfrak{T}$, *dans lesquels y est la seule variable à figurer*, deviennent, dès lors qu'on raisonne avec $\mathfrak{T}'$ et avec un de ses modèles M' obtenu, c'est évident, en complétant M par fixation d'un objet du domaine partiel à représenter par y dans $\mathfrak{T}'$, respectivement une relation close et un terme clos *de* $\mathfrak{T}'$; lorsque T est un terme de cette espèce, la vérification de la condition $(T1x)A$ dépend essentiellement du modèle M', et nous conclurons, par un raisonnement semblable, qu'il nous faudra, pour *tout* terme T de $\mathfrak{T}^\circ$ *dans lequel y est la seule variable à figurer*, la relation bien formée $(T^\circ 1x)A$ comme théorème de $\mathfrak{T}'$. D'un autre côté, sitôt cette condition remplie pour tout terme figurant dans une relation bien formée R de $\mathfrak{T}^\circ$ dans laquelle y est la seule variable à figurer, il n'est pas nécessaire d'aller au-delà de $\mathfrak{T}'$ pour se placer dans une théorie où sont satisfaites toutes les conditions dont nous disposons pour exprimer que chacun de ces termes désigne un objet du domaine partiel; on prendra donc $\mathfrak{T}'$ pour $\mathfrak{T}''$, et on aura à tenir une relation bien formée R de cette espèce pour valide dans le modèle M°, toutes les fois que son interprétation R° est un théorème de $\mathfrak{T}'$.

Par une récurrence évidente, on arrive à des conclusions analogues pour les assemblages bien formés comportant un nombre quelconque de variables. Etant donné un tel assemblage a, soit $\mathfrak{T}_a$ la théorie résultant de $\mathfrak{T}$ par adjonction de toutes les hypothèses $(y1x)A$ correspondant aux variables y de $\mathfrak{T}^\circ$ qui figurant dans a; notons qu'en vertu d'une remarque faite plus haut, $\mathfrak{T}_{a^\circ}$ est identique à $\mathfrak{T}_{a^\circ}$. En bref, nous en sommes au point suivant:

– pour réaliser ce qui a été décidé, *il faut faire en sorte* que l'interprétation d'un terme quelconque T de $\mathfrak{T}^\circ$ soit telle que $(T^\circ 1x)A$ soit un *theorème de* $\mathfrak{T}_T$;

– nous savons que toute relation bien formée R de $\mathfrak{T}^\circ$ dont *l'interprétation* R° est un *théorème de* $\mathfrak{T}_R$ est à coup sûr *valide dans le modèle* M°.

Le critère de la déduction permet de se débarrasser des théories $\mathfrak{T}_a$. Etant donné un assemblage bien formé a où figure au moins une variable de $\mathfrak{T}^\circ$, soit A_a la *conjonction* de toutes les relations bien formées $(y1x)A$ correspondant aux variables y de $\mathfrak{T}^\circ$ qui figurent dans a; les conditions auxquelles nous venons de parvenir *équivalent* aux suivantes (A_{a° étant aussi toujours la même que A_a):

– faire en sorte que, pour tout terme non clos (resp. clos) T de $\mathfrak{T}^\circ$, la relation bien formée $A_T \Rightarrow (T^\circ 1x)A$ (resp. $(T^\circ 1x)A$) soit un *théorème de* $\mathfrak{T}$;

– savoir pouvoir tenir une relation bien formée non close (resp. close) R de $\mathfrak{T}^\circ$ pour *valide dans le modèle* M°, lorsque la relation bien formée $A_R \Rightarrow R^\circ$ (resp. R°), qu'on désignera dans la suite par R^*, est un *théorème de* $\mathfrak{T}$.

Nous avons vu que R° apparaît comme représentant dans $\mathfrak{T}$ *la nature du lien* que la relation représentée dans $\mathfrak{T}^\circ$ par R établit entre les objets du modèle M°; par contre, c'est R^* qui, en indiquant de surcroît à quels objets du domaine d'objets de M se rapporte ce lien, apparaît comme *traduisant*, dans le langage de $\mathfrak{T}$, *le sens complet* de la relation bien formée R, *prise en tant que représentant une relation liant les objets d'un certain domaine, implicite dans* $\mathfrak{T}^\circ$, *et que cette traduction explicite, à sa manière.* Il sera donc naturel d'appeler R^* *la traduction* de la relation bien formée R; de ce point de vue, *le modèle* M° *se caractérise* par le *procédé de traduction* dont les règles formelles précises pour le passage de R à R^* ont été explicitées ou le seront dans la suite. C'est d'alleurs R^* qui *transpose de* $\mathfrak{T}^\circ$ *à* $\mathfrak{T}$ *la fonction remplie par R en tant que chaînon éventuel d'une démonstration;* on montre en effet sans peine que, *si R et R* $\Rightarrow$ *S sont deux relations bien formées de* $\mathfrak{T}^\circ$ *dont les traductions sont des théorèmes de* $\mathfrak{T}$, *il en va de même de S.* En conséquence, *si les traductions de tous les axiomes de* $\mathfrak{T}^\circ$ *sont des théorèmes de* $\mathfrak{T}$, *la traduction de tout théorème de* $\mathfrak{T}^\circ$ *sera un théorème de* $\mathfrak{T}$*;* or, ceci résulte immédiatement, pour les exemples des schémas $S1$, $S2$, $S3$, et $S4$ de ce qu'on sait déjà touchant leurs interprétations.[1])

La question se pose maintenant de savoir si le même cas se présente pour $S5$. Ceci amène à se demander ce qui se passe dans l'hypothèse où une relation bien formée $(\exists y)R$ est valide dans le modèle M°; or, si, tenant compte de ce qui précède, en ajoute à $\mathfrak{T}_{\tau_y(R)}$ l'hypothèse $((\exists y)R)^\circ$ – c'est-à-dire $((\tau_y(R))^\circ 1 y)R^\circ$, ainsi qu'on l'a vu –comme, si l'on a procédé correctement, $((\tau_y(R))^\circ 1 x)A$ est un théorème de $\mathfrak{T}_{\tau_y(R)}$, la relation bien formée $(\exists y)((y 1 x)A$ et $R^\circ)$ devient un théorème. Donc, de toutes façons, $((\exists y)R)^\circ \Rightarrow (\exists y)((y 1 x)A$ et $R^\circ)$ *est un théorème de* $\mathfrak{T}_{\tau_y(R)}$; donc, si $(\exists y)R$ est valide dans le modèle M°, la relation bien formée $(\exists y)((y 1 x)A$ et $R^\circ)$ est valide dans *tout* modèle $M_{\tau_y(R)}$ déduit de M par fixation, pour chacune des variables z de $\mathfrak{T}^\circ$ figurant dans $\tau_y(R)$, d'un objet du domaine partiel représenté par z dans $\mathfrak{T}_{\tau_y(R)}$.

[1]) L'ensemble des R^* pour R par courant l'ensemble des relations bien formées de $\mathfrak{T}^\circ$ constitue donc ce que I. L. Novak appelle dans (5) un *modèle réel* de $\mathfrak{T}^\circ$.

Sur le plan intuitif, si l'on accorde aux symboles le sens qu'on leur accorde habituellement, comme nous ne cessons de le faire ici, la situation est donc la suivante: si R est une relation bien formée de $\mathfrak{T}^\circ$, la relation de M qui a été prise dans le modèle M° pour relation 'il existe un objet vérifiant la relation représentée par R' *implique* la relation de M 'il existe un objet de M° vérifiant la relation représentée par R°' (qui, justement, a été prise pour relation représentée par R dans le modèle M°), lorsque les variables figurant dans les relations bien formées qui représentent ces relations de M désignent aussi des objets de M°. Ce fait s'accorde parfaitement avec la réalisation correcte de ce qui a été décidé; celle-ci cependant n'exige pas seulement cette implication, elle demande l'*équivalence* entre ces deux relations qui doivent être *vraies en même temps*, chacune d'elles affirmant l'existence d'un objet de M° vérifiant la relation de M prise pour relation de M° représentée dans $\mathfrak{T}^\circ$ par R; la première est la détermination, au moyen d'une relation de M, de la relation de M° affirmant cette existence; la seconde est la relation de M affirmant cette existence.

On peut réaliser cette condition en prenant pour $((\exists y)R)^\circ$ la relation bien formée $(\exists y)((y 1 x)A$ et $R^\circ)$ elle-même; c'est en cela que consiste la technique de relativisation habituelle en présence de quantificateurs primitifs. Mais c'est du symbole τ de Hilbert que nous disposons à ce titre, et pour réaliser ce qui a été décidé, il faudra et il suffira de faire en sorte que, pour *toute* relation bien formée R de $\mathfrak{T}^\circ$, la relation bien formée $(\exists y)((y 1 x)A$ et $R^\circ) \Rightarrow ((\tau_y(R))^\circ 1 y)R^\circ$ soit un théorème de $\mathfrak{T}_{\tau_y(R)}$. A titre de corollaire, si R est une relation bien formée de $\mathfrak{T}^\circ$ telle que le terme $\tau_y(R)$ est non clos (resp. clos), les relations bien formées $((\exists y)R)^*$ et $A_{\tau_y(R)} \Rightarrow (\exists y)((y 1 x)A$ et $R^\circ)$ (resp. $(\exists y)((y 1 x)A$ et $R^\circ))$ seront alors *équivalentes dans* $\mathfrak{T}$, ce qui fera admettre l'abus de langage d'appeler *traduction* de $(\exists y)R$ la seconde de ces relations, dont l'emploi revient à raisonner directement par usage des propriétés qu'on vient de requérir ou d'établir.

Considérons maintenant un exemple $(T 1 y)R \Rightarrow (\exists y)R$ de $S5$; si l'on a procédé correctement, $(T^\circ 1 y)R^\circ \Rightarrow (\exists y)((y 1 x)A$ et $R^\circ)$ est un théorème de $\mathfrak{T}_{(T1y)R \Rightarrow (\exists y)R}$; avec ce qu'on vient de voir, le critère de la déduction est juste ce qu'il nous faut pour en déduire que *tout exemple de S5 a aussi pour traduction un théorème de* $\mathfrak{T}$.

Nous avons donc établi que *si, pour tout terme T de* $\mathfrak{T}^\circ$, *la relation bien formée* $(T^\circ 1x)A$ *est un théorème de* $\mathfrak{T}_T$, *et si, pour toute relation bien formée R de* $\mathfrak{T}^\circ$, *la relation bien formée* $(\exists y)((y1x)A$ et $R^\circ) \Rightarrow ((\tau_y(R))^\circ 1y)$ R° *est un théorème de* $\mathfrak{T}_{\tau_y(R)}$, *il suffit, pour que tout théorème de* $\mathfrak{T}^\circ$ *soit effectivement valide dans le modèle* M°, *que tout axiome* spécifique *de* $\mathfrak{T}^\circ$ *ait pour traduction un théorème de* $\mathfrak{T}$ (Métathéorème I).

Il faut dire un mot du cas où $\mathfrak{T}$ et $\mathfrak{T}^\circ$ sont toutes deux *égalitaires*. Ce qu'on peut faire alors de plus simple en matière d'interprétation est de prendre $=$ pour $=^\circ$; on dira dans ce cas que le modèle M° est *équiégalitaire à* $\mathfrak{T}$.

Si tel est le cas, l'interprétation d'un exemple quelconque de *S*6 est encore un exemple de *S*6, et donc un théorème de $\mathfrak{T}$; *tout exemple de S*6 *a donc pour traduction un théorème de* $\mathfrak{T}$, et par voie de conséquence, *les exemples de S*6 *ne sont pas à compter au nombre des axiomes spécifiques de* $\mathfrak{T}^\circ$.

Il en va de même des exemples de S7, lorsque, en outre, A est un théorème de $\mathfrak{T}$.

En effet, sous ces conditions, comportant le seul résidu syntactique possible du fait que *A* est valide dans *M*, tous les développements qui précèdent se simplifient, ainsi que leurs conclusions, dont certaines sont en outre alors trivialement satisfaites; on voit sans peine qu'il suffit de montrer que $(\forall y)(R^\circ \Leftrightarrow S^\circ) \Rightarrow (\tau_y(R))^\circ = (\tau_y(S))^\circ$ est un théorème de $\mathfrak{T}$, quelles que soient les relations bien formées *R* et *S* de $\mathfrak{T}^\circ$; soit alors $\tau_y(\Sigma(R^\circ))$ l'interprétation $(\tau_y(R))^\circ$ de $\tau_y(R)$, où $\Sigma(P)$ désigne le schéma de relation attaché au modèle M° dont on a déjà parlé et dont la donnée de la relation bien formée *P* achève de déterminer un exemple; l'hypothèse supplémentaire $(\forall y)(R^\circ \Leftrightarrow S^\circ)$ fait de $(\forall y)(\Sigma(R^\circ) \Leftrightarrow \Sigma(S^\circ))$ un théorème, et le résultat souhaité résulte de *S*7 et d'une application du critère de la déduction.

En tout cas, lorsque $\mathfrak{T}^\circ$ est égalitaire, et que M° *n'est pas* équiégalitaire à $\mathfrak{T}$ (ce peut être le cas, simplement parceque $\mathfrak{T}$ *n'est pas* égalitaire), *il est nécessaire de vérifier que les exemples de S6 et de S7 ont pour traductions des théorèmes de* $\mathfrak{T}$; ces exemples doivent alors être regardés comme des *axiomes spécifiques* de $\mathfrak{T}^\circ$.

Nous nous bornons maintenant à l'examen du cas où $\mathfrak{T}$ *est égalitaire*, et où *l'on prend pour interprétation de* $\tau_y(R)$ *le terme* $\tau_y((y1x)A$ et

$\{(\exists y)((y1x)A$ et $R^\circ) \Rightarrow (R^\circ$ et $\Sigma(R^\circ))\})$, *où* $\Sigma(P)$ *désigne un certain schéma de relation dont la donnée de la relation bien formée* P *détermine complètement un exemple, et tel que, pour toute relation bien formée* R *de* $\mathfrak{T}^\circ$, *on ait* $(\exists y)((y1x)A$ et $R^\circ) \Rightarrow (\exists y)((y1x)A$ et R° et $\Sigma(R^\circ))$ *pour théorème de* $\mathfrak{T}_{\tau_y(R)}$.

Si ces conditions sont réalisées, les hypothèses du métathéorème I sont satisfaites; car, ou bien on a $\neg(\exists y)((y1x)A$ et $R^\circ)$, doù résultent $(\tau_y(R))^\circ = \tau_x(A)$, puis $((\tau_y(R))^\circ 1x)A$, puisque $(\exists x)A$ est un théorème de $\mathfrak{T}$; ou bien on a $(\exists y)((y1x)A$ et $R^\circ)$, d'où résultent $(\tau_y(R))^\circ = \tau_y((y1x)A$ et R° et $\Sigma(R^\circ))$, puis $((\tau_y(R))^\circ 1x)A$, d'une part, et $((\tau_y(R))^\circ 1y)R^\circ$, d'autre part, en utilisant la condition imposée dans l'énoncé. Donc, *ce cas particulier est un de ceux où tout théorème de* $\mathfrak{T}^\circ$ *est valide dans* M°, *pourvu que tout axiome spécifique de* $\mathfrak{T}^\circ$ *ait pour traduction un théorème de* $\mathfrak{T}$.

Si, de plus, $\mathfrak{T}^\circ$ *est égalitaire et* M° *équiégalitaire à* $\mathfrak{T}$, en cherchant à établir une démonstration analogue de ce que la traduction de tout exemple de *S7* est un théorème de $\mathfrak{T}$, on aboutit à la conclusion que tel est le cas, et donc que *tout exemple de S6 et de S7 est à ne pas compter au nombre des axiomes spécifiques de* $\mathfrak{T}^\circ$, *lorsque le schéma* $\Sigma(P)$ *aura été choisi de facon à ce que soit réalisée l'une des conditions suivantes:*

– *il existe un schéma* $\Sigma_1(P)$ *tel que, pour toute relation bien formée* R *de* $\mathfrak{T}^\circ$, *on ait* $(\exists y)((y1x)A$ et $R^\circ) \Rightarrow (\Sigma(R^\circ) \Leftrightarrow \Sigma_1((y1x)A$ et $R^\circ))$ *pour théorème de* $\mathfrak{T}_{\tau_y(R)}$;

– *il existe un schéma* $\Sigma_2(P)$ *tel que, pour toute relation bien formée* R *de* $\mathfrak{T}^\circ$, *on ait* $(\exists y)((y1x)A$ et $R^\circ) \Rightarrow \{((y1x)A$ et $\Sigma(R^\circ)) \Leftrightarrow \Sigma_2((y1x)A$ et $R^\circ)\}$ *pour théorème de* $\mathfrak{T}_{\tau_y(R)}$;

– *il existe un schéma* $\Sigma_3(P)$ *tel que, pour toute relation bien formée* R *de* $\mathfrak{T}^\circ$, *on ait* $(\exists y)((y1x)A$ et $R^\circ) \Rightarrow \{((y1x)A$ et R° et $\Sigma(R^\circ)) \Leftrightarrow \Sigma_3((y1x)A$ et $R^\circ)\}$ *pour théorème de* $\mathfrak{T}_{\tau_y(R)}$.

Le schéma $\Sigma_0(P)$ identique à $((y1x)A$ et $P) \Rightarrow ((y1x)A$ et $P)$ fournit un exemple de schéma $\Sigma(P)$ vérifiant, avec le schéma $\Sigma_1(P)$ identique à $P \Rightarrow P$, la première de ces trois conditions; il répond à toutes les autres conditions voulues à divers endroits. Or, tout exemple de $(P$ et $\Sigma_0(P)) \Leftrightarrow P$ est un théorème; donc, si c'est $\Sigma_0(P)$ qu'on emploie pour $\Sigma(P)$, l'égalité $(\tau_y(R))^\circ = \tau_y((y1x)A$ et $\{(\exists y)((y1x)A$ et $R^\circ) \Rightarrow R^\circ\})$ est un théorème de $\mathfrak{T}$, quelle que soit la relation bien formée R de $\mathfrak{T}^\circ$. Il en résulte aussitôt que, *si* $\mathfrak{T}$ *est égalitaire, et si l'on prend pour interprétation de* $\tau_y(R)$ *le terme* $\tau_y((y1x)A$ et $\{(\exists y)((y1x)A$ et $R^\circ) \Rightarrow R^\circ\})$, *tout théorème de* $\mathfrak{T}^\circ$ *est valide* *dans $M°$, pourvu que tous les axiomes spécifiques de $\mathfrak{T}°$, au nombre desquels il n'y a pas lieu de compter les exemples de S6 et de S7 si $\mathfrak{T}°$ est égalitaire et $M°$ équiégalitaire à $\mathfrak{T}$, aient pour traduction un théorème de $\mathfrak{T}$.* On comprendra ce qui se passe en remarquant que, dans ce cas, si l'on a $(\exists y)((y \mid x)A$ et $R°)$, on a aussi $(\tau_y(R))° = \tau_y((y \mid x)A$ et $R°)$. Or, il est clair qu'en s'en tenant à cela, on dispose de tout ce qu'il faut pour adapter au système de Bourbaki un grand nombre de démonstrations classiques, telles par exemple que celles de l'indépendance de l'axiome de l'infini et de la compatibilité de l'axiome de fondement avec les autres axiomes de la théorie des ensembles.

L'intérêt de la présence du schéma $\Sigma(P)$ au sein de celui qui s'attache au modèle $M°$ pour l'interprétation du τ, présence qui s'avère donc *facultative*, réside en ce qu'il restitue correctement, pour les quantificateurs définis du système de Bourbaki, la technique habituelle de relativisation, tout en accordant un peu plus: il peut permettre en outre l'étude de conditions auxquelles on veut soumettre la fonction logique de choix; j'ai pu ainsi démontrer la compatibilité, avec les autres axiomes de la théorie des ensembles de Bourbaki renforcée par l'hypothèse de l'existence de cardinaux inaccessibles, de certain *schéma* de ce genre, sur la portée duquel on s'était posé des questions. On peut aussi s'arranger dans certains cas pour choisir $\Sigma(P)$ de telle sorte que, pour toute relation R de $\mathfrak{T}°$ pour laquelle $(\exists y)R$ est valide dans le modèle $M°$, la relation bien formée '$(y \mid x)A$ et $R°$ et $\Sigma(R°)$' soit *fonctionnelle en y:* cela permet aux gens désireux de se rendre compte de ce que sont les êtres d'une théorie dont ils s'occupent d'en avoir un modèle au sein duquel tous les termes sont aisément situables.

C.N.R.S.

BIBLIOGRAPHIE

1. G. Asser – *Theorie der logischen Auswahlfunktionen*, Z. Math. Logik. u. Grundlagen Math. 3 (1957) pp. 30–68.
2. N. Bourbaki – Eléments de Mathématique. Livre 1 : Théorie des Ensembles. Chapitres I et II. Actualités scientifiques et industrielles, no. 1212, Hermann, Paris (1954).
3. D. Hilbert – Gesammelte Abhandlungen, t. III. Springer, Berlin (1935).
4. K. Gödel – The consistency of the continuum hypothesis, Princeton (1940).
5. I. L. Novak – *A construction for models of consistent systems*, Fundam. Math. 37 (1950), pp. 87–110.
6. A. Tarski, A. Mostowski, R. M. Robinson – Undecidables Theories, Studies in Logic, North-Holland Publ. Co., Amsterdam (1953).
7. H. Wang – *Arithmetic translations of axiom systems*, Trans. Amer. Math. Soc. 71 (1951), pp. 283–293.

A. KUIPERS

MODEL AND INSIGHT

I. INTRODUCTION

The contrast between modern conceptions of science and those of Aristotle is especially obvious in contemporary treatises on the method of natural science. We particularize this problem by trying to discover where there can be found points of similarity and dissimilarity between Aristotle's ideal of explanation and the factual explanations given in physics.

To Plato's dialectics the Stagirite opposes the search for truth, insight and certainty, which are proper to science in general. In form and content the scientific chain of reasoning must exclude any element of the will that is alien to science: the conclusion, therefore, will have to follow inevitably from the starting-point, and the starting-point must express essential features of the subjects at issue, which means an exact search of what the things are in themselves. The final premisses, which cannot be the result of reasoning, and also the principles of the demonstration, are obtained by an induction which is not so much a reasoning as a direct insight, an intuitive induction.

Scientific reasoning is restricted to what is necessary and to what is contingent in so far as the latter contains an element of necessity. Facts already known by sensual perception are digested by reason in an analysis which finally results in the knowledge that the phenomena have to be what they are and cannot be otherwise; the reasoning is therefore no anticipation of experience but a digestion of it, not leading to any new conclusions but to an insight into facts already known from experience. Scientific explanation is dependent on a knowledge of the essence of things, of what is in them by virtue of their nature. To penetrate and analyse this essence is the task of science; it does not come towards us spontaneously, but must be elicited from what the senses offer us.

During the time between Aristotle's days and ours, natural science has undergone a radical change, which may be characterized by the development of the experimental method. In the 16–17th century this new empirical method is established in its essentials and science begins to break away from philosophic reflection: science is no longer led by a philosophic view now, but by a method in which mathematics, as a means of thought and expression, begins to occupy a more and more prominent place. In the field of philosophy there arises, in addition to the classic Plato-Aristotle controversy, the controversy of rationalism versus empiricism and that of idealism versus realism. The philosophical systems diverge in the degree of emphasis they lay on a certain facet, on one central idea, dominating their entire cast of thought. Hence we try to solve the problem facing us by starting, not from a certain philosophic view, but from the factual physical explanation with all the aspects proper to it.

II. EXPLANATIONS OF CLASSICAL MECHANICS

A closer analysis of the explanations of classical mechanics shows that the enumerating of forces is coupled with an endeavour to reach a deeper background by which something of the nature of things is revealed. Nature, as seen by Aristotle, is found again as a functional relation between quantities which are regarded as being essential to the phenomenon. On the other hand there manifests itself a difference of opinion as regards the position of the experience which is completed by the experiment, and as regards the relativity of the explanation which originates in the use of a model and the shifting from one level to another.

The best way to define the function of the model seems to be to refer to the theory of fall. From this it will be clear how very intricate are the phenomena that we observe daily around us, and in addition to this, that science can offer us only a very simplified picture of that reality which we have abstracted from all kinds of disturbing circumstances.

There are many intermediate phases in passing from physical reality, in which a material body makes its way through a resisting medium, to the abstract wording where a material point moves through a vacuum. Consequently, the phenomenon is made from something very complicated and concrete into an ideality which is closely related to mathematical entities. In spite of all abstractions and idealisations the model remains a physical model, because all quantities such as mass, weight, velocity and acceleration are directly concerned with the concrete physical phenomenon. It is not mathematical but physical notions that define the picture given. There are however many points of connection between mathematics and this physical model.

We consider the falling body as a material point. In this way we know what route is taken by the centre of gravity, but we don't know the movements of this body round the centre of gravity. Also the material point, which plays such an important part in the abstract formulation, is concerned with the actual phenomenon, and is distinguished from the mathematical point by its mass. When however we have to describe the exact route taken by this material point, we shall have to consider the body as a mathematical point, whose place is represented in course of time by a certain function. If therefore we take one step further towards abstraction by discarding mass, then the problem has become a purely mathematical one.

Measuring offers us in fact two series, to which correspond two numbers: the length of the route and the time spent, or speed at given moment and the moment itself. The simplest way to see the functional relation between the quantities appearing here is a diagram. The curve which then appears seems to lead to a half parabola or a straight line. This is an attempt to arrange the results of mensuration in practical form, but we are also looking for a certain regularity, in the conviction that such a regularity exists. Once having found we take it as being of general avail. We therefore switch over from this concrete finite series of numbers to an arbitrary pair of numbers, in which one number may be derived from the other in the same way in which, in the concrete series, the observed quantities correspond. So we pass from a series of discrete values of the time measured to a collection of values which is continuous.

In these manipulations we discard the dimensions and the correspondence between the numbers and the phenomena. Thus the physical problem has been reduced to a mathematical one. We are now looking for a formal relation, a mathematical function, according to which the series may be connected. So we are looking for a mathematical model corresponding to the series of numbers which are the results of the physical measurement. The starting-point for this is to be found in the table in which the physical measurements are given. The result is a function predicting the distance covered in a given time.

The diagram is also a mathematical model of the law of gravitation.

The continuous line therefore represents the supposed regularity. The table had been constructed by means of an experiment in a vacuum. So the distance is technically limited within a maximum, unlike the mathematical function which is therefore an extrapolation.

III. OPTICAL PHENOMENA

In explaining optical phenomena the main accent falls on the use of different models for light, which occur in all kinds of varieties. In geometrical optics light is regarded as something travelling along straight lines. To this model corresponds a geometrical method of representation by which we can predict the phenomena that take place in reality.

New prescriptions – namely the equation for lenses – have to be added to the rules for the technique of representation, if we extend the field of application from formation of shadows to the refraction of light by a lens.

The explanation always remains partial. We can ask new questions to which this model offers no answer: What is that something travelling along straight lines? Is it particles or is it a wave? So we try to specify the model given, either in Newton's emission-theory or in Huygens's wave-theory, so as to explain refraction and reflection. We refer to phenomena that we can observe around us and for which we have already formed a theory, supposing that in the world of our daily experiences the same laws hold good as in the microcosm.

The equations of Maxwell offer a further specification of the wave-theory, representing them as periodical variations of electric and magnetic field-strengths. Thus the model of light has become a mathematical equation which will serve as a starting-point from which we derive the various properties of light. The quantities occurring in this equation are of a physical character, so that they cannot vary at will and in solving the equations we are tied to the actual situation which has induced us to form these equations.

The simplest model has preference; the physicist retains it until the phenomena force him to a further differentiation. Through this the explanations become more partial in character. This is not a feature of optical explanations: every explanation is essentially connected with the use of a certain model.

This results from the physicist's attitude of mind, which induces him to drop of the real phenomenon all that is not, or is not supposed to be, important for his measurement. The physical model is a link between the direct observations and the abstract mathematical theory, which knits these experiences logically together and thus explains them.

IV. MACROCOSM

If we are to know the route of the earth round the sun, we consider the earth as a mass-point; that is to say: we discard its dimensions and imagine its mass compressed into one point. An ellipse whose large and small axes have the same proportions as those of the earth's route gives us a picture of the successive distances of the earth in relation to the sun. It is a model of the movement of the earth round the sun, in which we see, in reduced form, what happens in reality.

This reminds us of the models used in shipbuilding, where the actual situation is imitated as faithfully as possible. There is, however, an important difference between this ellipse-model and the models used in shipbuilding. In both cases we speak of the actual situation and compare the model with it. In the case of shipbuilding we find ourselves in a field in which we can survey that situation by means of our senses, in contrast with the case of the macrocosm, in which our observation fails. Of the sun we only know the gold and yellow circle and of the earth we only survey the patch of ground within our circle of vision. When we speak of the distance between the sun and the earth the human mind has purposely detached itself from the insufficient subjective experiences and placed itself on a standpoint from which it can survey the situation. We imagine that we are standing in the cosmos at a point where we can see the earth with one eye and the sun with the other, just as we are used to do with the objects in our immediate surroundings. So when speaking of the relation between the ellipse-model and the actual situation we mean the relation between an imagined situation and two mass-points whose distance varies with time.

Apart from this the earth is also considered as a body permanent in form and dimensions. The problem is quite a different one now; we are not trying to define the distance from the earth to the sun as a function of time, but we are looking for the movements made by the earth itself, the movements round the centre of gravity, the rotation of the earth and the phenomena connected with this. To understand and explain the Coriolis-forces it is sufficient to introduce the model of a merry-go-round or that of a rotating globe. If we want to make evident the fact that the acceleration of gravitation depends on the place on the earth, we shall have to turn to the deformation of the earth and to replace the globe as model by the ellipsoid.

When however we have to explain the deformation of the earth, then we arrive at a field where we have to consider the fluid state which the earth once had. We can no longer consider it as a fixed body with the rotation of a top. Our attention is mainly fixed on the fluidity of the body, which when at rest as a result of gravitation-forces assumes the shape of a globe, but which, in consequence of its rotation round a fixed axis, is flattened by the influence of centrifugal forces.

So we have three essentially different models of the earth: the material point, the ellipsoid and the fluid model. These relate to three different problems: the movement of the centre of gravity, the movement round the centre of gravity and the deformation of the earth. These three models cannot be reduced to one another, although they relate to the same real body. They are artificial constructions, which have to be thought out after observing the phenomena and which at the same time must have some sense in connection with the problem on hand. This requires equivalence between the model and what is pictured – equivalence in those aspects, which are of importance for this particular problem. Whether we can directly ascertain this equivalence depends on the level at which the phenomena take place. The physical model never pretends to be a faithful copy of reality; only the main lines, with which we are concerned here, in this particular problem, must be sharply outlined in the model, so that we can survey the whole, and thanks to this survey get an insight into the real state of affairs.

The term equivalence in this connection means that the quantitative relations which measurement shows to prevail in the real situation should prevail for the model as well. The representative force of the model is especially and exclusively to be found in its correspondence with the real phenomenon as regards the quantitative relations. It is immaterial how these relations are realised in the actual situation.

V. MICROPHYSICS

The question which is becoming increasingly prominent in microphysics is the following: What is it in the microcosm that corresponds to the phenomena that we can state on our level? How are we to picture to ourselves the structure of the atom, in order to be able to deduce the perceptible effects? So we desire a model that corresponds to the phenomena that we can observe and that point in the direction of the interior of the atom.

There also has to be a correspondence with phenomena already explained and arising from the interior of the atom. That is why Bohr starts from the Rutherford atom-model and modifies it so as to make it agree with the line spectrum.

Last of all we require the model to correspond with the laws and models known to us from macrophysics. The way in which we approach the microcosm is quite dependent on our available knowledge. These are in the first place our experiences in our own world. For everything outside the field of our immediate sensual perceptions we use the experiences of our own world as model.

From the macrophysical world we know laws and models whose applicability and fertility had already been proved. We have discovered them by extrapolating our ordinary experiences to the field that our senses cannot survey. The success of this passage from our own experience into the domain of the cosmos via the model causes us to suppose that the same laws will hold good in microcosm. Whether this supposition is correct will have to appear from its results. Thus the representation of the atom as a miniature solar system is in fact a model that proved helpful in understanding the laws of the macrocosm, and therefore comes to function as a model in the microcosm.

In constructing modern physics we notice a shifting from what may be concretely pictured to what is mathematically abstract. Physics in its most modern shape finds a gradually decreasing number of points of connection with the imaginitive faculty of man, and the symbols of physical formulation cannot always be brought into correlation with physical quantities. Classical physics uses models which are illustrations built up from known elements. Physical theory in quantum-mechanics is gradually moving away from the perceptible picture. As we penetrate further, expression becomes more abstract; concrete representations are decreasingly suitable to represent the structure of matter. Mathematical formalism on the other hand gradually increases in importance. As however the developing theory always arises as a result of a generalisation of the preceding theory, there remains a connection with the physical model, which in each case forms the point of connection for building up the formalism by means of which we try to describe the situation. The mathematical model has acquired the upper hand; the physical model we can only use to give a concrete representation of a few facets of the formalism.

When once the formalism has been founded and its ability has been proved, we might discard every physical model by putting forward the formalism with its rules, which fix the way in which we have to handle the symbols used, and by giving prescriptions bringing about a correlation between certain symbolised quantities and the results of measuring apparatus. All this however under the supposition that the formalism is complete and that there is no need to extend it. If however we want to add to the formalism or extend its field of application, then we shall have to turn again to the physical model, which is a link between formalism and physical reality.

VI. CONCLUSION

Together with the model a cardinal point of difference has been indicated between Aristotle's ideal and the factual course of things in physics.

In the model we find at the same time the points of similarity and dissimilarity between Aristotle's doctrine and the factual physical explanations. By means of the model we try to grasp the essential facets of a physical process; on the other hand, the genesis of a model involves a living contact with sensual observation. The model does not originate spontaneously in the human mind, but requires creative activity. Thus senses and intellect both play an active part in our shaping of the model and consequently in our obtaining an insight into the phenomena which cause us to try and find explanations.

O.E.S.A., Haarlem

J. J. MULCKHUYSE

MOLECULES AND MODELS [1)]

1.0. INTRODUCTION

Most of the studies in axiomatization and formalization of theories from natural science are about theories of physics and biology. The theory that is discussed here is from the field of chemistry. It is a rather typical theory of the old stamp, that is to say, developed in a time before the merging of physics and chemistry: the structure theory of Couper, Kekulé, van 't Hoff and Le Bel.

It is a compliment to the intuition of the founders of this theory, that it is still to-day a very important tool in the hands of the organic chemist although the defects and limitations of this theory are now very well known.

The fundamental facts of this theory are still valid and can be derived from a more fundamental theory: wave-mechanics, as a special case.

2.0. DISCUSSION OF THE CONTENTS OF STRUCTURAL CONSIDERATIONS IN ORGANIC CHEMISTRY

In general most textbooks of organic chemistry give discussions about the structure of molecules which are a mixture of at least three theories, that is to say, the structure theory of Kekulé and van 't Hoff the electronic theory of the chemical bond of Lewis and the wave-mechanical theory (Pauling).

The most fundamental concept in all these discussions is that of the valency of an atom: that is the number that indicates the amount of hydrogen atoms that can be bound by an atom or an atomgroup, or the number of hydrogenatoms the atom or group of atoms can replace. The valency of hydrogen is one.

1) This article is part of a study about the structure theory in organic chemistry which was presented as a thesis.

The aim of the structure theory proper was perhaps best formulated by van 't Hoff:
'The structure theory studies the fixed positions of the atom in a molecule in respect to each other in space according to geometrical principles without considering the nature of the binding forces.'
That this geometrical representation has always remained the main purpose of this theory is proved by the modern development of it, the so called Stuart models of molecules. These models are a result of the refinements that have been introduced in the structure theory since van 't Hoff. These Stuart molecule models are made of wooden or plastic balls representing atoms and metal connections representing the chemical bonds, observing the rules for distances between the atoms, angles between the bonds etc. punctually.
It became, however, evident that such a geometrical theory was not capable of explaining all known facts. Therefore Lewis designed an electronic theory. In this theory the bonds are represented by two electrons. The special merit was that it was possible to give a qualitative descriptions of ionic structures which could not be described by the geometrical theory. But it was not before wave-mechanics made its bow that a better physical description could be given for chemical phenomena and in our case for chemical structures.
The present situation is a rather peculiar one. On the one hand the new wave mechanical concepts proved to be too difficult to cope with the daily work of the organic chemist so that the old geometrical theory remained in use, on the other hand wave mechanics corroborated nearly all the suppositions of the geometrical structuretheory such as the tetrahedral building of the carbonatom in a molecule. Therefore this geometric theory is still in full use with regard to organic chemistry and with great success.
This fact seems to justify an attempt at an axiomatic description of this theory. For a general description of the facts of the structure theory we can refer to textbooks on organic chemistry like Fieser and Fieser and textbooks about the chemical bond like Pauling and Ketelaar.
In this article we will give an axiomatization and formalization making use of elementary logic only. See for a further discussion with the use of the second-order logic: Mulckhuyse (1).

3.0. THE AXIOMATIZATION AND FORMALIZATION OF THE STRUCTURE THEORY

3.1. *Structure theory and species of atoms*

The species of atoms, considered in organic chemistry and to which the structure theory is applicable, are numerous. First of all we have carbon atoms followed by hydrogen, oxygen, nitrogen, sulphur, fluorine, chlorine, bromine, iodine, silicium, phosphorus, boron and some metals. Because, however, the essential facts of the structure theory can already be adequately explained, if use is made of two species only namely carbon and hydrogen, only these two kinds of atoms will be introduced. Nothing fundamental is gained by introduction of other species.

Another important point is that the structure theory gives models for physical entities, which are called molecules. One model stands for myriads of molecules as an example. And a third point to be mentioned is that the structure theory imposes no limitations upon the size of a molecule. The models, that can be constructed, will be of unlimited size although each model is determined by its empirical formula: C_nH_m or $C_nH_mO_r$ or $C_nH_mO_rN_s$. ('empirical molecular formula' will be abbreviated as 'EMF') [1]).

3.2. *Individuals, sets and relations*

The axiomatization can be clearly divided in two parts: an abstract part (theory A), which deals only with sets of individuals and their relations and a geometrical part (theory A′), which deals with the spatial distribution of the individuals, and can be considered as an extension of the abstract theory. The geometrical theory is obtained from the abstract theory (i) by introducing a representation of the individuals and their relations in space, and (ij) by adding several empirical data, which are necessary for understanding certain properties of a molecule.

The first step in axiomatizing the abstract part A (which is meant to deal as we know with a set of individuals and certain relations) is the

1) On thermodynamical considerations it is however evident that the size of for instance a carbon atom chain is not unlimited. As soon as the size of the chain is so that the amount of kinetic energy per atom of 3/2 kT, if added together, is greater than the bondenergy of a C-C bond the probability, that the chain will break, is greater than zero.

introduction of a system $\mathfrak{A} = \langle \mathscr{A}, \mathscr{C}, \mathscr{H} \rangle$ composed of sets $\mathscr{A}$, $\mathscr{C}$ and $\mathscr{H}$ of individuals.

Individuals: The atoms of the EMF are represented in the system $\mathfrak{A}$ by individuals; the variables $x_1, x_2, \ldots, x_k, \ldots$ range over the individuals of the sets of system $\mathfrak{A}$.

Sets: These individuals form a set $\mathscr{A}$.

The following subsets are discussed:

$\mathscr{C}$, consisting of all individuals, which represent carbon atoms,

$\mathscr{H}$, consisting of all individuals, which represent hydrogen atoms.

It is sometimes convenient to name the individuals of the subsets independently. This will be done by the constants c and h:

$$\mathscr{C} = \{c_1, c_2, \ldots, c_n\}$$
$$\mathscr{H} = \{h_1, h_2, \ldots, h_m\}$$

Relations: Because the individuals of $\mathscr{A}$ are connected with each other in different ways, binary relations r_1, r_2, and r_3 are introduced; accordingly we shall need atomic formulas:

$$r_1(x_1, x_2),\ r_2(x_1, x_2),\ r_3(x_1, x_2).$$

This means that an individual x_1 is connected with an individual x_2 of $\mathscr{A}$ by the relations r_1, r_2, or r_3. Or more specifically:

r_1 expresses that the individual x_1 is connected with x_2 by a single bond;

r_2 expresses that these individuals are connected by a double bond;

r_3 expresses that they are connected by a triple bond.

For the present moment we have thus the systems $\mathfrak{A} = \langle \mathscr{A}, \mathscr{C}, \mathscr{H} \rangle$.

Well-formed formulas: If x_k and x_l are individual variables of $\mathscr{A}$ the following formulas are well-formed formulas (wff's).

(i) $x_k \varepsilon \mathscr{A}$, $x_k \varepsilon \mathscr{C}$, $x_k \varepsilon \mathscr{H}$.

(ij) $r_1(x_k, x_l)$, $r_2(x_k, x_l)$, $r_3(x_k, x_l)$.

(iij) $x_k = x_l$

(iv) all formulas formed from (i)-(iij) by the connective signs of elementary logic.

(v) no other formulas than those mentioned in (i)-(iv) are well-formed.

3.3. *General axioms on individuals and relations (theory A)*

First some properties of sets $\mathscr{A}$, $\mathscr{C}$ and $\mathscr{H}$ will be formulated. These steps in the axiomatization can be done with only elementary logic as a tool (at a certain point however it will be necessary to appeal to second-order logic so as to give an exhaustive description of the structure theory).

The first two axioms are:

(A.1) The set $\mathscr{A}$ is built up only of individuals of $\mathscr{C}$ and $\mathscr{H}$ and contains no other individuals.

$$(x)(x \,\varepsilon\, \mathscr{A} . \rightarrow . \, x \,\varepsilon\, \mathscr{C} \vee x \,\varepsilon\, \mathscr{H})$$

(A.2) The sets $\mathscr{C}$ and $\mathscr{H}$ have no individuals in common

$$\overline{(\mathrm{E}x)(x \,\varepsilon\, \mathscr{C} \;\&\; x \,\varepsilon\, \mathscr{H})}$$

And the two properties, that characterize the r relations, are given in the next two axioms:

(A.3) The relations r_1, r_2 and r_3 are irreflexive (no individual is connected with itself) and mutually exclusive.

$$(x_k)\,[\overline{r_1(x_k, x_k)} \;\&\; \overline{r_2(x_k, x_k)} \;\&\; \overline{r_3(x_k, x_k)} \;\&$$
$$\& \; (x_l)\,\{\overline{r_1(x_k, x_l) \;\& r_2(x_k, x_l)} \;\& \overline{r_1(x_k, x_l) \;\&\; r_3(x_k, x_l)} \;\&$$
$$\& \overline{r_2(x_k, x_l) \;\&\; r_3(x_k, x_1)}\}]$$

(A.4) The relations r_1, r_2 and r_3 are symmetric (if an individual x_k is connected with an individual x_l, then x_l also with x_k)

$$(x_k)(x_l)\,\{r_1(x_k, x_l) \rightarrow r_l(x_l, x_k) \;.\; \& \;.\; r_2(x_k, x_l) \rightarrow r_2(x_l, x_k) \;.\; \& \;.$$
$$r_3(x_k, x_l) \rightarrow r_3(x_l, x_k)\}.$$

To sum up: the above relations are:

(a) irreflexive,
(b) exclusive,
(c) symmetric,
(d) neither transitive, nor intransitive,
(e) neither strongly connected, nor even connected in the set $\mathscr{A}$.

The next two axioms deal with the properties of individuals of the subsets $\mathscr{C}$ and $\mathscr{H}$.

(A.5) An individual x of $\mathscr{C}$ can be argument:

(i) four times in the r_1-relation, or
(ij) twice in the r_1-relation and once in the r_2-relation, or
(iij) twice in the r_2-relation, or
(iv) once in the r_1-relation and once in the r_3-relation.
(v) These are the only possibilities for an individual of $\mathscr{C}$.

$$(x_1)\,[x_1 \varepsilon \mathscr{C} \rightarrow \{(Ex_2)(Ex_3)(Ex_4)(Ex_5)\ [x_2 \neq x_3\ \&\ x_2 \neq x_4\ \&$$
$$\&\ x_2 \neq x_5\ \&\ x_3 \neq x_4\ \&\ x_3 \neq x_5\ \&\ x_4 \neq x_5\ \&$$
$$\&(x_6)\ \{r_1(x_1, x_6) \leftrightarrow (x_6 = x_2 \vee x_6 = x_3 \vee x_6 = x_4 \vee$$
$$x_6 = x_5)\}]\} \vee \{(Ex_2)(Ex_3)(Ex_4)[x_2 \neq x_3\ \&\ x_2 \neq x_4\ \&\ x_3 \neq x_4\ \&$$
$$\&\ (x_5)\ \{r_1(x_1, x_5) \leftrightarrow (x_5 = x_2 \vee x_5 = x_3)\ .\ \&\ .\ r_2\ (x_1, x_5) \leftrightarrow$$
$$x_5 = x_4\}]\} \vee$$
$$\{(Ex_2)(Ex_3)\ [x_2 \neq x_3\ \&\ (x_4)\ \{(r_2(x_1, x_4) \leftrightarrow x_4 = x_2 \vee$$
$$x_4 = x_3) \vee (r_1(x_1, x_4) \leftrightarrow x_4 = x_2\ .\ \&\ .\ r_3(x_1, x_4) \leftrightarrow x_4 = x_3)\}]\}].$$

(A.6) An individual of $\mathscr{H}$ can be argument only once in the r_1-relation.

$$(x_1)\,[x_1 \varepsilon \mathscr{H} \rightarrow \{(Ex_2)(x_3)\ [r_1(x_1, x_3) \leftrightarrow x_2 = x_3]\}].$$

It is now possible to construct from a given system $\mathfrak{A} = \langle \mathscr{A}, \mathscr{C}, \mathscr{H} \rangle$ a number of models: $\mathfrak{S} = \langle \mathscr{A}, \mathscr{C}, \mathscr{H}, r_1, r_2, r_3 \rangle$ called abstract structures, making use of the axioms (A.1)-(A.6).

It is however not yet possible to characterize those structures, which can be considered as abstract representations of the molecules of organic chemistry. The concept that is necessary here is that of coherence; all the individuals of set $\mathscr{A}$ must be connected with each other by the relations r_1, r_2 or r_3 for a construction of a given system $\mathfrak{S}$. It is however impossible to define this concept in elementary logic without imposing certain restrictions on our theory A. For the definition first an arbitrary finite number of individuals must be considered, and secondly it is necessary to speak of properties of arbitrary abstract structures. This difficulty can be overcome by confining axiomatically the number of individuals to a finite amount, say 136.2^{256}. Because we do not want to give such an axiom here, we prefer to formulate a rule, which forms no part of our formalized theory A. As a shorthand notation we introduce here $R(x_1, x_2)$ which expresses that an individual x_1 is connected with an individual x_2 by the relation r_1, r_2 or r_3.

Rule of coherence: A model of axiom system A is called coherent if for all x_1 and for all x_2 there is a finite sequence of individuals $y_1, y_2, \ldots, y_n$ so that x_1 is identical with y_1 and x_2 with y_n and $R(y_1, y_2)\ \&\ R(y_2, y_3)\ \&\ \ldots$ $\&\ R(y_{n-1}, y_n)$ is valid. Only coherent models are taken into consideration.

[If we restrict our system $\mathfrak{A}$ by an axiom like:

There are at most 136.2^{256} individuals in set $\mathscr{A}$,

it is possible to incorporate this concept of coherence in our theory A (as Suppes pointed out) by introduction of an axiom scheme. We first have to define the shorthand notation suggested above:

(d1) $R'(x_1, x_2) \leftrightarrow [r_1(x_1, x_2) \vee r_2(x_1, x_2) \vee r_3(x_1, x_2)]$.

The axiom scheme, that can be formulated now, has the general form:

$$(Ex_1)(Ex_2)\ldots(Ex_k)\, P_k(x_1, x_2, \ldots, x_k) \rightarrow$$
$$(Ex_1)(Ex_2)\ldots(Ex_k)\, R_k(x_1, x_2, \ldots, x_k),$$

where the predicates P_k and R_k are defined as follows:

(1) $S_{k+1}(x_1, x_2, \ldots, x_k, y)$ is $S_k(x_1, x_2, \ldots, x_{k-1}, y)\ \&\ x_k \neq y$.

(2) $P_{k+1}(x_1, x_2, \ldots, x_{k+1})$ is $P_k(x_1, x_2, \ldots, x_k)\, S_{k+1}(x_1, x_2, \ldots, x_k, x_{k+1})$.

(3) $T_{k+1}(x_1, x_2, \ldots, x_k, y)$ is $T_k(x_1, x_2, \ldots, x_{k-1}) \vee R'(x_k, y)$.

(4) $R_{k+1}(x_1, x_2, \ldots, x_{k+1})$ is $T_{k+1}(x_1, x_2, \ldots, x_{k+1})\ \&\ R_k(x_1, x_2, \ldots, x_k)$.

and

(1) $S_2(x_1, x_2)$	is $x_1 \neq x_2$,
(2) $P_2(x_1, x_2)$	is $x_1 \neq x_2$,
(3) $T_2(x_1, x_2)$	is $R'(x_1, x_2)$,
(4) $R_1(x_1, x_2)$	is $R'(x_1, x_2)$.

As an example take a case of 4 individuals:

$S_4(x_1, x_2, x_3, x_4)$	is $S_3(x_1, x_2, x_4)\ \&\ x_3 \neq x_4$
$S_3(x_1, x_2, x_4)$	is $x_1 \neq x_4\ \&\ x_2 \neq x_4$.
$P_4(x_1, x_2, x_3, x_4)$	is $P_3(x_1, x_2, x_3)\ \&\ S_4(x_1, x_2, x_3, x_4)$
$P_3(x_1, x_2, x_3)$	is $x_1 \neq x_2\ \&\ S_3(x_1, x_2, x_3)$
$S_3(x_1, x_2, x_3)$	is $x_1 \neq x_3\ \&\ x_2 \neq x_3$
$T_4(x_1, x_2, x_3, x_4)$	is $T_3(x_1, x_2, x_4) \vee R'(x_3, x_4)$
$T_3(x_1, x_2, x_4)$	is $R'(x_1, x_4) \vee R'(x_2, x_4)$
$R_4(x_1, x_2, x_3, x_4)$	is $T_4(x_1, x_2, x_3, x_4)\ \&\ R_3(x_1, x_2, x_3)$
$R_3(x_1, x_2, x_3)$	is $T_3(x_1, x_2, x_3)\ \&\ R'(x_1, x_2)$
$T_3(x_1, x_2, x_3)$	is $R'(x_1, x_3) \vee R'(x_2, x_3)$

and thus: $(Ex_1)(Ex_2)(Ex_3)(Ex_4)\{x_1 \neq x_2\ \&\ x_1 \neq x_3\ \&\ x_2 \neq x_3\ \&\ x_1 \neq x_4\ \&\ x_2 \neq x_4\ \&\ x_3 \neq x_4\} \rightarrow (Ex_1)(Ex_2)(Ex_3)(Ex_4)\{R'(x_1, x_4) \wedge \vee R'(x_2, x_4) \vee R'(x_3, x_4).\&.R'(x_1, x_3) \vee R'(x_2, x_3).\&.R'(x_1, x_2)\}$.

3.4. *Construction of models*

The axioms (A1)-(A6) form a first elementary theory A. We shall now consider the generation of models $\mathfrak{S}$ from this theory. We have called the models over a given system $\mathfrak{A}$ abstract structures. As known already there are a great number of models, which do not have a physical meaning: these incoherent models will be discarded.

The abstract structures can be generated in several ways. Perhaps the most elegant way is to make use of the matrices, which give an immediate overall-picture of all the possibilities. For the generation of such matrices a machine can be constructed. The following rules of construction can be given.

Rules of construction:

(i) Form all possible matrices over the axioms (A1)-(A6) for a given system $\mathfrak{A}$ for the individuals $c_1, \ldots, c_n, h_1, \ldots, h_m$ with respect to the relation in which these individuals can be arguments.

This can be done because for a given system $\mathfrak{A}$ there is only a finite number of possibilities. Each matrix represents an abstract structure.

(ij) Select only those matrices whose abstract structures are in accordance with the rule of coherence.

It follows immediately from (A3) that only matrices, which have zero's in the main diagonal, are generated. Further it is evident that only matrices that are symmetric in respect to this diagonal are generated according to (A4) and (A5). Further only those matrices are generated whose columns contain for every individual exactly the relations that are allowed by (A5)-(A6).

In the example that is given now some possible matrices are developed for a given EMF.

Example. The given EMF is C_3H_6.

The system is $\mathfrak{A} = \langle \mathscr{A}, \mathscr{C}, \mathscr{H} \rangle = \langle \{c_1, c_2, c_3, h_1, h_2, h_3, h_4, h_5, h_6\} \rangle$.

First step: form the matrices according to rule (i). (only a few matrices are given here).

1.

	c_1	c_2	c_3	h_1	h_2	h_3	h_4	h_5	h_6
c_1	0	r_1	r_1	r_1	r_1	0	0	0	0
c_2	r_1	0	r_1	0	0	r_1	r_1	0	0
c_3	r_1	r_1	0	0	0	0	0	r_1	r_1
h_1	r_1	0	0	0	0	0	0	0	0
h_2	r_1	0	0	0	0	0	0	0	0
h_3	0	r_1	0	0	0	0	0	0	0
h_4	0	r_1	0	0	0	0	0	0	0
h_5	0	0	r_1	0	0	0	0	0	0
h_6	0	0	r_1	0	0	0	0	0	0

2.

	c_1	c_2	c_3	h_1	h_2	h_3	h_4	h_5	h_6
c_1	0	r_2	0	r_1	r_1	0	0	0	0
c_2	r_2	0	r_2	0	0	0	0	0	0
c_3	0	r_2	0	0	0	r_1	r_1	0	0
h_1	r_1	0	0	0	0	0	0	0	0
h_2	r_1	0	0	0	0	0	0	0	0
h_3	0	0	r_1	0	0	0	0	0	0
h_4	0	0	r_1	0	0	0	0	0	0
h_5	0	0	0	0	0	0	0	0	r_1
h_6	0	0	0	0	0	0	0	r_1	0

To the matrices 1 and 2 correspond the following chemical formulas:

1.

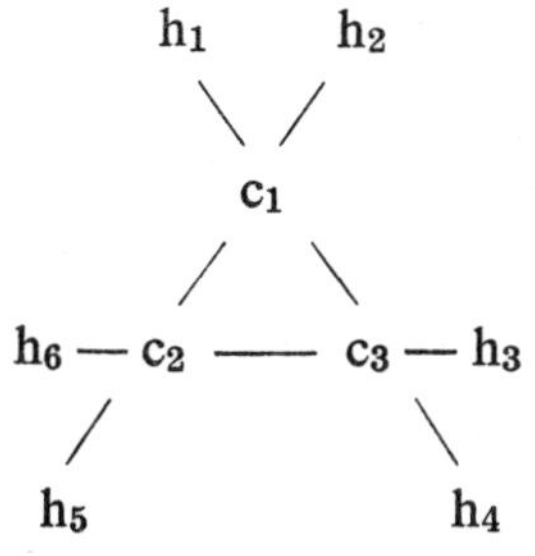

2.

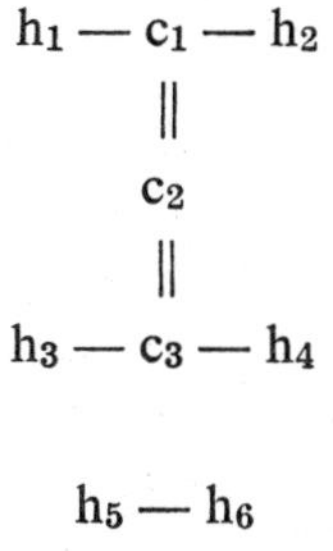

Second step: Applying rule (ij) reject those matrices which produce incoherent models. This is here matrix 2. Only 1 remains as a valid model.

3.5. *Spatial representation of the axiom system (A.1)-(A.6): Theory A'*

The purely abstract axiomatization given in the foregoing sections can be built out a little further in an abstract way, but it is impossible to give a complete abstract axiom system for the structure theory, because there are typical spatial properties, which can be expressed in the best way by making use of geometrical notions.

From a formal point of view it is necessary to use an axiomatized geometry for the investigation of the geometrical representations of the abstract structures. It is of course not allowed to use an intuitive geometry. Because we want to restrict the logical tools to first-order logic an axiomatization of geometry in this logic has to be chosen. A very good one has been given by Tarski. This system is based on two primitive notions:

a) the ternary precidate β, denoting the betweenness relation;

b) the quarternary predicate δ, denoting the equidistance relation.

The two dimensional system as given by Tarski has to be developed into a three dimensional or solid geometry. As Tarski, p. 21 remarks all the results in his paper can be adapted to the three dimensional case. The axioms 11 and 12 of Tarski have to be modified, leaving the other axioms unchanged.

By giving definitions of lines etc. in terms of the predicates β and δ it is possible to develop the greater part of Euclidian geometry using only points as individuals (see Tarski).

Now the set $\mathscr{A}$ is mapped onto a set $\mathscr{A}'$ of points in space. The function which establishes this representation is denoted as ψ_2. This ψ_2 has to be applied to the individuals and relations of the systems $\mathfrak{S}$. The relations r_1, r_2 and r_3 are then represented by relations r'_1, r'_2 and r'_3. The line segment d with the points p_1 and p_2 as endpoints, is denoted by $d(p_1p_2)$. Any other line will be denoted by reference to two of its points, for instance (ab). An important characteristic of the chemical bond can now be expressed namely its length. With each pair of points p_1, p_2 we associate in addition to the line segment $d(p_1p_2)$ a non-negative real number μ, called the distance between p_1 and p_2.

The following rule allows the application of the function ψ_2 to the abstract structures.

Representation rule

(i) the individuals of $\mathscr{A}$ are mapped by ψ_2 on points:

$$\psi_2(x_1) = p_1;$$

(ij) the relations r_1, r_2 and r_3 are represented by the spatial relations r'_1, r'_2 and r'_3

$$(x_1)(x_2)(Ep_1)(Ep_2)\{\psi_2(x_1) = p_1 \;\&\; \psi_2(x_2) = p_2 \;\&\; (r_1(x_1, x_2) \rightarrow r'_1(p_1, p_2)) \;\&\; (r_2(x_1, x_2) \rightarrow r'_2(p_1, p_2)) \;\&\; (r_3(x_1, x_2) \rightarrow r'_3(p_1, p_2))\}.$$

To characterize the geometrical property that is associated with the r'-relations namely the length of a line segment, we introduce the following rule which associates a line segment of definite length with each well-formed formula $r'_1(p_1, p_2)$, $r'_2(p_1, p_2)$, $r'_3(p_1, p_2)$.

Bond length rule

(i) $(p_1)(p_2)\,[p_1 \,\varepsilon\, \mathscr{C}' \;\&\; p_2 \,\varepsilon\, \mathscr{C}' \;\&\; (r'_1(p_1, p_2) \rightarrow d(p_1p_2) = \mu_1) \;\&\; (r'_2(p_1, p_2) \rightarrow d(p_1p_2) = \mu_2) \;\&\; (r'_3(p_1, p_2) = \mu_3)]$.

(ij) $(p_1)(p_2)\,[p_1 \,\varepsilon\, \mathscr{C}' \;\&\; p_2 \,\varepsilon\, \mathscr{H}' \;\&\; (r'(p_1, p_2) \rightarrow d(p_1p_2) = \mu_4)]$.

The values μ_1, μ_2, μ_3 and μ_4 are for our purpose quite irrelevant. The best values at the moment are resp. 1,54 Å, 1,22 Å, and 1,09 Å.

This mapping in space of the abstract structures, that are formed in accordance with the construction rule, the bond length rule and axioms (A1)-(A6), results in an infinite number of realizations for each given abstract structure: there are of course an infinite number of ways of mapping an individual *b* which has for instance the relation $r_1(a, b)$ to an individual a when a has already been mapped. This is however inessential: all these mappings of the first relation are identical, the direction is in this case not essential. But if two individuals *a* and *b* are mapped which have for instance the relation $r_1(a, b)$ whereas $\psi_2(a) = a'$ and $\psi_2(b) = b'$ and if a' and b' are now connected by the line segment $d(a'b')$, then the number of possibilities of mapping an individual *c* which has for instance the relation $r_1(b, c)$ to *b*, is again infinite.

And that means, that segment $d(b'c')$ can have an infinite number of positions with respect to segment $d(a'b')$ which directly contradicts the facts of the structure theory. It is necessary to adapt the axiom system

so as to provide for new experimental facts, thus the bond angles are introduced. It will be clear that then only a few mappings result in acceptable models.
These models will be called spatial structures of molecules according to general use.
We must emphasize once more, to prevent misunderstanding, that we have thus two kinds of structures. The *abstract structures* as models of the theory A, which could be generated in several ways; we have chosen for this purpose the matrix-form. These abstract structures must be distinguished from the *spatial structures*, which are models of our theory A′. These spatial structures are, as already said, of a geometrical nature.

3.6. *Introduction of the bond angles. Bond model for carbon*

The angle φ between two line segments is usually considered as a ternary function of three points *a*, *b*, and *c*: $\varphi(a, b, c)$. However, because a notation from which we can see directly by which line segments the angle is formed is much more practical, the angle function will be written as a binary function of the line segments. This is especially convenient if we wish to denote angles between crossing line segments. Thus $\varphi(d(ab), d(cd))$ is the angle formed by the line segments $d(ab)$ and $d(cd)$; for intersecting line segments $b = c$.
The original system $\mathfrak{A} = \langle \mathscr{A}, \mathscr{C}, \mathscr{H} \rangle$ is now replaced by:

$$\mathfrak{A}' = \langle \mathscr{A}', \mathscr{C}', \mathscr{H}' \rangle$$

and the system

$$\mathfrak{S} = \langle \mathscr{A}, \mathscr{C}, \mathscr{H}, r_1, r_2, r_3 \rangle$$

by the system

$$\mathfrak{S}' = \langle \mathscr{A}', \mathscr{C}', \mathscr{H}', \psi_2, r'_1, r'_2, r'_3, \varphi \rangle.$$

The bond models for carbon can be introduced in several ways. The most elegant one is the classical way of van 't Hoff and Le Bel, who introduced the tetrahedron bondmodel. As known a tetrahedron is a rectilinaer figure determined by four not collinear points. These points are the vertices of the tetrahedron, the line segments between these points are the sides and the triangles formed by the sides the faces of the tetrahedron.
(d3) The point $p_1 \varepsilon \mathscr{A}'$, has the property T with respect to the points

p_2, p_3, p_4 and p_5 ε $\mathscr{A}'$ (T(p_1; p_2, p_3, p_4, p_5)) if there is a tetrahedron with the points p_2, p_3, p_4 and p_5 as vertices whereas the angles formed by the line segments $d(p_1p_2)$, $d(p_1p_3)$, $d(p_1p_4)$, $d(p_1p_5)$ are all equal (and hence given by $\sin a = {}^2/_3 \sqrt{6}$).

This tetrahedron model as a bond model for the carbon atom is only applicable if the carbon atom is bound by four single bonds. Although this model can be adapted for the case of a carbon atom bound by two single and one double bond, this is rather artificial: in this case we must postulate in the first place that the double bond is situated on the bisector of the angle formed by two line segments (p_1p_2) and (p_1p_3) and in the second place that the two line segments make an angle of 120° with each other. Therefore we prefer to introduce here another bond model.

(d4) The point p_1 ε $\mathscr{A}'$, has the property D wich respect to the points p_2, p_3 and p_4 ε $\mathscr{A}'$ (D(p_1; p_2 p_3, p_4)) if there is a triangle with the points p_2, p_3 and p_4 as vertices whereas the angles formed by the line segments $d(p_1p_2)$, $d(p_1p_3)$ and $d(p_1p_4)$ are all equal and the points p_1, p_2. p_3 and p_4 are coplanar.

And for the case of the atom that is bound by a triple and a single bond or by two double bonds:

(d5) The point p_1 ε $\mathscr{A}'$ has the property L with respect to the points p_2 and p_3 ε $\mathscr{A}'$ (L(p_1; p_2, p_3)) whereas p_1, p_2 and p_3 are collinear.

We can now formulate the axiom, which postulates the different bond models for the case of a carbon atom and which we have called

Van 't Hoff axiom.

(A'7) Every individual p_1 ε $\mathscr{C}'$ has,
(i) if p_1 is four times argument in the r'_1-relation, the property T;
(ij) if p_1 is twice argument in the r'_1-relation and once in the r'_2-relation, the property D;
(iij) if p_1 is twice argument in the r'_2-relation, or once in the r'_1-relation and once in the r'_3-relation, the property L.

$(p_1)\ [p_1\ \varepsilon\ \mathscr{C}' \rightarrow$
$(Ep_2)(Ep_3)(Ep_4)(Ep_5)\ \{r'_1(p_1p_2)\ \&\ \ldots\ \&\ r'_1(p_1p_5) \rightarrow$
$T(p_1; p_2, p_3, p_4, p_5)\}$
$\vee\ (Ep_2)(Ep_3)(Ep_4)\ \{r'_1(p_1p_2)\ \&\ r'_1(p_1p_3)\ \&\ r'_2(p_1p_4) \rightarrow$
$D(p_1; p_2, p_3, p_4)\}$

$\vee\ (Ep_2)(Ep_3)\ \{(r'_2(p_1p_2)\ \&\ r'_2(p_1p_3))\vee(r'_1(p_1p_2)\ \&\ r'_3(p_1p_3))\rightarrow L(p_1;\ p_2, p_3)\}].$

The theoretical values for the bond angles given in the definition (d3), (d4) and (d5) are not always exactly in accordance with the experimental facts, which is ascribed to a small influence of the groups, which are connected with the carbon atom in question.
This dependence is here neglected as is generally done in structural considerations.

3.7. *Free rotation and cis-trans isomerism*

'Since all molecular models which are interconvertible by rotation of the centers of gravity of the atoms about their common axis correspond to one and the same chemical individual, the auxiliary hypothesis which must be added (to the structure theory), as a sort of set of instructions for the use of the space-molecule models in order to explain the number of isomers, is known as the hypothesis of so-called 'free' rotation'. (Hückel)
This rotation is, however, not really free: for normal hydrocarbons there is a potential barrier of 3 kcal/mol which has to be surmounted (see especially Ketelaar p. 208 etc.). A rotation along the double and triple bonds is, however, impossible.
The possibility of rotation along a double bond must be excluded explicitly from our system, because it is otherwise not possible to speak about the so-called cis-trans isomers.
This can be performed in the following way. If for two points p_1 and $p_2\ \varepsilon\ \mathscr{C}'$ and the points p_3, p_4, p_5 and $p_6\ \varepsilon\ \mathscr{A}'$ and $r'_2(p_1, p_2)$, $r'_1(p_1, p_3)$, $r'_1(p_1, p_4)$, $r'_1(p_2, p_5)$ and $r'_1(p_2, p_6)$ and if the planes of the triangles belonging to the points p_1 and p_2 are called respectively $P(p_1;\ p_2, p_3, p_4)$ and $P(p_2;\ p_1, p_5, p_6)$, the following axiom is introduced:

$$(A'8)\ (p_1)(p_2)(p_3)(p_4)(p_5)(p_6)\ [r'_2(p_1, p_2)\ \&\ D(p_1;p_2, p_3, p_4)\ \&\ D(p_2;\ p_1, p_4, p_6)\rightarrow P(p_1;\ p_2, p_3, p_4) = P(p_2;\ p_1, p_4, p_6)].$$

A direct consequence of this axiom is that if we have $r'_2(p_1, p_2)$ and further $r'_1(p_1, p_3)$, $r'_1(p_1, p_4)$, $r'_1(p_2, p_5)$ and $r'(p_2, p_6)$, then the point p_3 can have two positions in respect to point p_5, which two positions are not identical:

(Theorem) If $p_1\ \varepsilon\ \mathscr{C}'$ and $p_2\ \varepsilon\ \mathscr{C}'$ and $r'_2(p_1, p_2)$, $r'_1(p_1, p_3)$, $r'_1(p_1, p_4)$, $r'_1(p_2, p_5)$ and $r'_1(p_2, p_6)$, then $\varphi\ (d(p_1p_3), d(p_2p_4))$ can have two values only: π or $^1/_3\,\pi$.

The proof follows immediately from (A′8) and definition (d4).
From this theorem follows the possibility of cis-trans isomerism (cis if φ $(d(p_1p_3), d(p_2p_4)) = {}^1/_3\, \pi$, trans if $\varphi = \pi$).
If for the points p_1, p_2, p_3 and p_4 we have $r'_3(p_1, p_2)$, $r'_1(p_2, p_3)$ and $r'_1(p_2, p_4)$, then the points p_3, p_1, p_2, p_4 are collinear (see (A′7)) and a rotation along $d(p_1p_2)$ has no effect on the position of points p_3 and p_4 to each other. A spatial axiom for this case is thus not given.
The axiom system A′ permits us to discuss all the special structures of hydrocarbon chemistry with a few exceptions only. The whole system as given here is based on elementary logic. However, if we want to discuss the facts of stereo-isomerism, it will be necessary to make use of a second-order logic. This discussion is given in Mulckhuyse.

3.8. *Limitations of the axiom system A′*

In several respects the axiom system A′ is deficient. On one side it is possible to generate point-line structures whose real counterparts, the molecules, can not exist; on the other side there exists a great number of molecules whose structures can not be deduced from the given axiom system. Nevertheless this system is the crystallization of the ideas that are the base of the classical structure theory as developed by Kekulé, van 't Hoff and Le Bel.
The defects of the axiom system are due to the fact that this classical theory is deficient itself. These limitations can be divided in five groups:
(1) Limitations on the level of our abstract theory A.
There are some cases of stable radicals. The structure of these radicals can not be deduced from our system.
(2) Limitations on the level of the geometrical theory A′.
All cases where there are steric hindrances, which can be explained by the introduction of atom radii, so that 'free' rotation can not exist or bond-angles are changed.
(3) Limitations of a dynamical character.
(i) The existence of some ringstructures is not consistent with our theory. Well known cases are cyclopropane and cyclobutane. Von Baeyer introduced his strain hypothesis in which he supposed that the value of the valence angles can vary between certain limits and thus does not have a fixed value.
(ij) Some substances can be considered as an equilibrium between two

structures (to be distinguished from resonance). Our theory does not take into account such an equilibrium state.
An example is the equilibrium state between the boat- and chair-structures of cyclohexane.
(4) Limitations due to ionized structures. Several substances can be considered as partly ionized. It is evident that new assumptions will have to be made to adapt our theory to these experimental facts. Especially when nitrogen and oxygen atoms are introduced the introduction of ionized structures will be necessary.
(5) Wave mechanical limitations:
(i) All cases where the resonance method has to be explicitly applied to give satisfying results such as the aromatic ringstructures and the conjugated double bond. In general the physical facts can best be explained by postulating a resonance between several covalent bond structures and ionized structures (which do not exist separately). The benzene molecule can be taken as an example.
(ij) in our theory A and A′ the assumption was made that the four bonds of a carbon atom are equal in properties. Wave-mechanical considerations, however, make this assumption doubtful. As is known the best solution of the wave equation is a combination of the four functions furnished by one 2s and three 2p-functions of the carbon atom in the ^5S-state.
This combinations result in all probability in one function, which is directed from the centre to the vertices of a tetrahedron, just as postulated by van 't Hoff in 1874. These four functions which are a result of the combination of the 2s and 2p-functions, are not exactly directed to the vertices of a regular tetrahedron, however, if the four groups bound to the carbon atom are different.
For a discussion of these limitations and several extensions of our axiomsystem A and A′ which remove part of the limitations we refer to Mulckhuyse.

4.0. SOME PROPERTIES OF THE AXIOM SYSTEM A: INDEPENDENCE, COMPLETENESS AND DECIDABILITY

The *independence* of the axiom (A1)-(A6) is easily established by using models of this system. It is easy to see that the axioms, which establish the properties of subsets $\mathscr{C}$ and $\mathscr{H}$ of set $\mathscr{A}$ and of irreflexivity and of symmetry of the relations r_1, r_2 and r_3 are independent.

The independence of axiom (A5) can be shown by taking a model, which satisfies (A1)-(A4) and (A6) but not (A5).
Take for instance the model which is given in the form of a matrix for

$$\mathscr{A} = \{c_1, c_2, h_1, h_2, h_3, h_4\}$$

	c_1	c_2	h_1	h_2	h_3	h_4
c_1	0	r_3	r_1	r_1	0	0
c_2	r_3	0	0	0	r_1	r_1
h_1	r_1	0	0	0	0	0
h_2	r_1	0	0	0	0	0
h_3	0	r_1	0	0	0	0
h_4	0	r_1	0	0	0	0

This model satisfies axioms (A1)-(A4) and (A6) but not (A5) because this axiom says that an individual of $\mathscr{C}$ can be at most once argument in the r_3-relation and once in the r_1-relation.
The model as given here should agree with a molecule $H_2C \equiv CH_2$, which does not exist.
The same kind of argument is valid for the independence of (A6).
About the completeness of A we can make the following observations. As known a theory A is called incomplete if there is a statement U without free variables, which can be formulated in the language of A but can not be proved in A while its negation can not be proved either. It is very easy to give such a statement: there is an individual $x_1 \varepsilon \mathscr{C}$ that is connected with four individuals of $\mathscr{H}$ and no other individuals exist. Or formalized:

$$\text{(E1)}\ (\mathrm{E}x_1)(\mathrm{E}x_2)(\mathrm{E}x_3)(\mathrm{E}x_4)(\mathrm{E}x_5)\left[x_1 \varepsilon \mathscr{C} \mathbin{\&} x_2 \varepsilon \mathscr{H} \mathbin{\&} x_3 \varepsilon \mathscr{H} \mathbin{\&} x_4 \varepsilon \mathscr{H} \mathbin{\&} \right.$$
$$\mathbin{\&} x_5 \varepsilon \mathscr{H} \mathbin{\&} r_1(x_1, x_2) \mathbin{\&} r_1(x_1, x_3) \mathbin{\&} r_1(x_1, x_4) \mathbin{\&} r_1(x_1, x_5) \mathbin{\&}$$
$$\left. \mathbin{\&} x_1 \neq x_2 \mathbin{\&} \ldots \mathbin{\&} x_4 \neq x_5 \mathbin{\&} (x_6)\{x_6 = x_1 \vee \ldots \vee x_6 = x_5\}\right].$$

This statement describes a molecule CH_4. It is evident that it is impossible to derive (E1) or its negation from our axioms (A1)-(A6) (CH_4 as described by (E1) is however a model of our axiom system).
If we could derive (E1) from our axioms then all the models of these axioms would also be models of (E1) and therefore the structure described by the statement (E1) would be the only one permitted by our axioms. As however other structures as C_2H_6 are also allowed for by

axioms (A1)-(A6), it follows that (E1) is not derivable from these axioms.

If the negation of (E1) could be derived (or in other words: if (E1) would be inconsistent with our axioms), then these axioms would not permit the structure CH_4 and therefore the negation can not be either derived from them.

Thus our theory A is incomplete. This form of incompleteness is of course necessary for our purpose, because otherwise our theory would have only *one* model (in the case of (E1) as a derivable statement: the methane structure). However, this desirable incompleteness has another consequence. Because we have a theory which has finite models having any finite number of individuals, also infinite models can be generated (see e.g. Beth (1) p. 555).

These infinite models are not interesting from a chemical point of view. If we limit the axiom system to one given system $\mathfrak{A}$ these infinite models disappear. A given system $\mathfrak{A}$ has a finite set of individuals $\mathscr{A}$ and therefore it is only possible to give a certain number of finite models. This is a weak kind of completeness.

The foregoing discussion led us onto the following considerations.

We can make extensions of axiom system A that are complete. To attain this we add to the axiom system an expression that characterizes a single molecule for instance as given in (E1). The axiom system (A1)-(A6) & (E1) is complete, it has only one model. (If it had two models a new extension can be made which only has one model).

We have made the assumption that this extension is finitely axiomatisable. If this is not the case there is no finite model, only an infinite model which is, as already said, not interesting from a chemical point of view. We can thus state that all the complete extensions which generate chemically interesting structures are finitely axiomatisable. The following reduction can now no made. Every closed expression is concerned with (i) certain objects: individuals, like $x_1 \varepsilon \mathscr{C}$, and so on, and (ij) certain relations between two individuals: r_1, r_2, r_3 or $\overline{r_1}$, $\overline{r_2}$, $\overline{r_3}$. Substitute now for every w.f.f. $r_1(x_1, x_2)$ etc. $x_1 \neq x_2$, etc. If there is no relation r_1, r_2 or r_3 between two individuals, they can be identical or non identical. The w.f.f.'s $x_1 \varepsilon \mathscr{C}$, and so on, are neglected. We get then an expression that says something about the number of individuals and nothing more. For instance (E1) becomes:

$$(E2)\ (Ex_1)(Ex_2)(Ex_3)(Ex_4)(Ex_5)$$
$$[x_1 \neq x_2 \ \& \ldots \& \ x_4 \neq x_5 \ \& \ (x_6) \{x_6 = x_1 \vee \ \ldots \ \vee x_6 = x_5\}]$$

Thus (E2) is an expression that characterizes the number of individuals namely 5. If we neglect now all the stereoisomers (which is evident because in the system A nothing can be said about these isomers) an interesting feature if this reduction is that all these finitely axiomatisable complete extensions determine all those structures, which have a meaning from a chemical point of view and that all the extensions that are not finitely axiomatisable generate structures that are chemically meaningless.

It is further rather likely that our theory A is *undecidable* although no attempt has been made to prove it.

Municipal University, Amsterdam

REFERENCES

Beth E. W., The foundations of mathematics. Amsterdam, 1959.
Fieser L. F. and Fieser M., Organic chemistry. New York, 1956.
Hoff J. H. van 't, Die Lagerung der Atome im Raume. Braunschweig, 1894.
Hückel W. and Rathmann F. H., Theoretical principles of organic chemistry. Vol. I. Amsterdam, 1955.
Ketelaar J. A. A., Chemical constitution. Amsterdam, 1958.
Lewis G. N., Valence and the structure of atoms and molecules. New York, 1923.
Mulckhuyse J. J., Molecules and models, investigations on the axiomatization of structure theory in chemistry. Amsterdam, Thesis, 1960.
Pauling L., The nature of the chemical bond. New York, 1945.
Suppes P., Personal communication.
Tarski A., What is elementary geometry. In Henkin L., Suppes P. and Tarski A., The axiomatic method. Amsterdam, 1959.

MODÈLE D'INTERACTION ENTRE CORPUSCULES EN THÉORIE FONCTIONNELLE

La Microphysique a-t-elle besoin d'un modèle pour se développer. Autrement dit, dans la description microscopique des phénomènes, quel est le rôle des modèles physiques?

Dès le début du développement de la Théorie des quanta on a proposé la représentation ponctuelle et le modèle planétaire de l'atome avec des schémas vectoriels de l'espace physique à trois dimensions dérivés des images classiques. La Mécanique ondulatoire s'est développée ensuite selon un schéma vectoriel dans l'espace de Hilbert. Cela veut dire que les connaissances acquises sur les particules sont représentées par des fonctions d'ondes qui sont des scalaires (particule de Klein-Gordon), des spineurs (théorie de Dirac), des vecteurs (photon et méson vectoriel de de Broglie), des tenseurs (champs de Maxwell quantifiés). Ces fonctions sont des éléments de l'espace vectoriel de Hilbert.

Ensuite avec Bohm, Vigier etc.... le modèle du fluide relativiste doté d'une agitation chaotique perpétuelle s'est formé. Les particules sont considérées comme un ensemble de tourbillons qui forment un fluide doté d'une densité de moment angulaire interne. Si l'image intuitive facilite l'étude des phénomènes microphysiques, par contre elle limite notre idée sur la réalité des choses.

Enfin le modèle fonctionnel des corpuscules a été proposé par Destouches.[1]) Le caractère objectif de la fonction *u* représentant la particule n'est pas intuitif mais il a l'intérêt d'être le plus adéquat parmi les modèles déjà proposés. Dans l'état actuel des théories, un modèle intuitif au sens classique du terme a peu de chances d'être adéquat et il faut recourir à des modèles abstraits offrant plus de directions possibles pour des développements. On pourra associer à la fonction physique *u* l'image ponctuelle ou le fluide continu du corpuscule par certaines fonctionnelles de *u*. La différence entre le modèle vectoriel de la Mécanique ondulatoire et le modèle fonctionnel de la théorie fonctionnelle est que les quantités

[1]) J. L. Destonches. *Corpuscules et champs en théorie fonctionnelle* (Gauthiers-Villars, Paris 1958).

liées aux connaissances et aux prévisions dans le premier cas s'expriment dans le schéma vectoriel de l'espace de Hilbert tandis que l'autre peut être exprimé dans un espace plus large. En plus l'interprétation physique de ces deux théories n'est pas la même puisque la Mécanique ondulatoire est purement probabiliste tandis que dans cette dernière théorie il y a à la fois un schéma probabiliste et une description objective.

MODÈLE D'INTERACTION ENTRE CORPUSCULES

Considérons un système de corpuscules $C_1\ C_2\ \ldots\ C_n$ dans lequel le j^e corpuscule C_j est représenté par une fonction u_j; chacune de ces fonctions u_j obéit à une équation d'évolution $L_j u_j = Q_j$. L'interaction entre ces corpuscules s'exprime par le principe que la forme des opérateurs L_j et des termes non linéaires Q_j dépend des autres fonctions u_k représentant les autres corpuscules. On voit apparaître ainsi une modification par rapport aux principes d'interaction entre corpuscules des autres théories. Soit un système d'un grand nombre N de particules de *même espèce* indépendantes l'une de l'autre. Chaque particule u_j obéit à une équation $L_j u_j = Q_j$. L'indépendance entre ces particules fait penser que les L_j ne dépendent pas de l'indice j, seuls les Q_j sont différents pour représenter l'objectivité des u_j. Dans ce cas, considérons la fonction moyenne $u_M = \frac{1}{N} \sum_i u_i$ de ces corpuscules indépendants.[1]) La fonction u_M représente sommairement le système pour étudier les influences de ce système sur le reste de l'Univers (en particulier sur les corpuscules de l'autre espèce): au lieu de sommer les interactions individuelles de chaque corpuscule C_j du système sur le reste de l'Univers on peut les calculer à l'aide de la fonction u_M. En plus les propriétés de u_M nous informe globalement sur l'état du système. Quant à l'étude séparée d'une particule C_j du système, il faut toujours utiliser l'équation $L_j u_j = Q_j$.

Sous quatres conditions [2]), u_M obéit à une équation linéaire $L u_M = 0$, ce qui nous conduit à penser que u_M se comporte, au point de vue prévisionnel comme l'onde ψ de la Mécanique ondulatoire. L'onde écart $u_{éj} = u_j - u_M$ obéit exactement à l'équation $L u_{éj} = Q_j$. de sorte

[1]) F. Aeschlimann. *J. Phys. Rad.* 20, 1959, p. 604.

[2]) J. L. Destouches et F. Aeschlimann, *Les systèmes de corpuscules en théorie fonctionnelle* (Hermann, Paris 1959).

que $u_{éj}$ se comporte comme l'onde aléatoire en théorie de la double solution, et le terme non linéaire Q_j est responsable de cette onde.
A titre d'exemple nous nous proposons d'étudier dans le modèle fonctionnel, les interactions entre deux classes de corpuscules bien connus: l'électron et le photon.[1]) Dans ce schéma, l'électron est représenté par une fonction u_e à quatre composantes qui obéit à un système de quatres équations aux dérivées partielles non linéaires que l'on écrit symboliquement par $L_e u_e = Q_e$. La fonction photonique u_p à seize composantes obéit à $L_p u_p = Q_p$. Les opérateurs linéaires L_e, L_p et les termes Q_e, Q_p dépendent respectivement des u_p et u_e.
L'interdépendance entre les L_e, L_p, Q_e, Q_p d'une part et les u_p, u_e d'autre part nous conduit à adopter l'idée que dans les équations électromagnétiques liées au photon il y a des termes dus à l'électron comme les moments électrique P, magnétique M; les densités de charge ρ, de courant i; par contre les champs créés par le photon agissent sur le mouvement de l'électron.
Si dans l'ancien modèle ponctuel, l'interaction entre corpuscules se manifeste par la section efficace, par l'image des chocs entre deux points matériels avec déviation de la trajectoire, dans le modèle fonctionnel l'interaction se manifeste par la contribution des propriétés intrinsèques des particules.

CONCLUSIONS

Il y a trop de distance entre les conceptions abstraites de la Physique et les modèles intuitifs empruntés à l'univers concret pour que l'on puisse espérer tirer de l'utilisation de ceux-ci des indications exactes. Il faut recourir à des modèles déjà conceptuels et abstraits empruntés plutôt à l'univers mathématique qu'à l'univers sensible.

Institut Henri Poincaré, Paris

[1]) Pham Xuân Yêm, *J. Phys. Rad.* Mars 1960.

J. F. STAAL

FORMAL STRUCTURES IN INDIAN LOGIC

There is a use of the term 'model' in which it can be said that a linguistic expression, in a natural language, is a model for its sense. A translation of a linguistic expression from one language into another may be said to provide another model for the sense of the original. If the sense of a linguistic expression is of a logical nature, the expression can be translated into an expression of formal logic or into a formula. This is not surprising, for logic and mathematics came into being when expressions of natural languages were translated into formal symbolisms, which were more precise and practical and less cumbersome. Subsequently these artificial languages attained full independence and started a development of their own. Originally, however, these symbolisms could only have been constructed along the lines suggested by the possibilities of expression and the scope of expression of the natural languages themselves. That in mathematics and in modern logic such a linguistic origin of the symbolism has often receded into the background does not imply that the origin of certain symbolisms was independent from the structure of natural languages [1]).

In view of this background it is not surprising that modern logic could provide the tools for the representation of logical expressions used by the Western logicians from Aristotle onwards. This has been shown by Bocheński, Łukasiewicz and many others. It is less evident, on the other hand, that the symbolism of modern logic should be useful in the representation of the only formal logic regarding which there are good reasons to believe that it developed independently from European logic: namely Indian logic. Nevertheless modern formalisms have been introduced – sometimes hesitatingly – into the study of Indian logic by

[1]) Some examples of such dependence are discussed in the present author's 'The construction of formal definitions of subject and predicate', to be published in *Transactions of the Philological Society*.

S.Sen [1]), S. Schayer [2]) and D. H. H. Ingalls [3]). I. M. Bocheński has now written the first comprehensive history of formal logic which takes Indian material into account [4]). That a modern symbolism can actually be used for the representation of Indian logic at all need not imply that this symbolism is necessarily universal. For Sanskrit, the language in which the Indian logicians expressed themselves even if their mother tongue was different, is an Indo-European language and its structure is largely similar to the structure of, for instance, Greek or Latin. This holds for its syntax as well as for its analysis of the parts of speech, both structures which are highly relevant for the development of a formal logic.

The origins of Indian logic are invisible, but the disciplines of reasoning developed in the speculations of later Vedic texts as well as in the researches of the Sanskrit grammarians. The famous grammar of Pāṇini (probably IVth century B.C.) reflects a very high level of logical reasoning and can only be considered as the fruit of a long developement, most of the traces of which are lost. The oldest logical text which has come down to us is the *Nyāya-sūtra*, which received its present form in the second or third century A.D. From then onwards an extensive logical literature was produced in India by Hindus, Buddhists and Jains. Logical techniques were adopted by some schools of philosophy and criticized and rejected by others. After a long period of logical discussions, in which the Buddhist logicians (e.g. Vasubandhu, Diṅnāga, Dharmakīrti) played a large part, a process of re-orientation took place between the Xth and the XIIth century. This culminated in the gigantic work of Gaṅgeśopādhyāya (Gaṅgeśa) (XIIIth century), founder of the 'New School' (*navya-nyāya*), when logic became largely free from philosophy, epistemology and cosmology, and the attention was mainly confined to the analysis of inference (*anumāna*). Logic thus became an instrument and a method, and as such it was used in various disciplines. Soon knowledge of the logical terminology and familiarity with the techniques of logical analysis became indispensable for anybody writing on matters philosophical, grammatical,

[1]) S. Sen, A study of Mathurānātha's *Tattva-cintāmaṇi-rahasya*. Wageningen, 1924.
[2]) S. Schayer, Über die Methode der Nyāya-Forschung. Festschrift M. Winternitz, Leipzig 1933, 247–57; and in other publications.
[3]) D. H. H. Ingalls, Materials for the study of Navya-nyāya logic. Cambridge, Mass. 1951. Cf. the present author's review in *Indo-Iranian Journal* 4 (1960), 68–73.
[4]) I. M. Bocheński, Formale Logik. Freiburg/München, 1956, 479–517; 'Die indische Gestalt der Logik'.

ritual and scientific in general. In the following centuries a new flow of logical literature was produced, mainly in Bengal in North-East India. Among the general handbooks then written, mention may be made of the *Siddhānta-muktāvalī* or *Kārikāvalī-muktāvalī* of Viśvanātha Pañcānana (XVIIth century) upon which the present study is based.

Indian logic has every right to be called formal from Gaṅgeśa, and possibly from the Buddhist logicians, onwards. It is formal in so far as it establishes formal rules, the validity of which depends on the structure of the sentence-expressions only. In such expressions variables occur (e.g. 'reason', 'conclusion') for which constants (e.g. 'smoke', 'fire') may be substituted. But while the presence and absence of such constants determine the validity of an empirical expression, they do not affect the validity of a logical expression.

The logical expressions are written in a kind of technical Sanskrit, where use is made of certain features of the Sanskrit language which lend themselves to a formalised treatment. Foremost among these features is nominal composition. As it is relevant in the present context to compare Sanskrit in this respect to other Indo-European languages, a recent formulation may be quoted: 'The capacity to combine independent words into compound words is inherited by Sanskrit from Indo-European, and similar formations are found in other IE languages. Sanskrit differs from the other IE languages in the enormous development which the system has undergone, which is unparelleled elsewhere'[1]).

We have elsewhere studied the relation between these linguistic means of expression and the logical structures [2]). The present paper is based upon a part of the material dealt with in that article, which mainly addresses readers who are familiar with Sanskrit. The present presentation is confined to a representation of Indian expressions by means of symbols and models of modern logic. For the Sanskrit originals the reader may be referred to the other article.

In the following, use is made of the terminology of the predicate calculus with equality, and in addition of the expression $axF(x)$ denoting the idea 'x such that $F(x)$'. We shall make use of the property: $(Ey)(y = axF(x)) \leftrightarrow (Ex)F(x)$. If there are several values of x such that $F(\dot{x})$, $axF(x)$ may

[1]) T. Burrow, The Sanskrit language. London, 1955, 207–8.

[2]) 'Correlations between language and logic in Indian thought', Bulletin of the School of Oriental and African Studies 23 (1960), 109–122.

denote any of these values: e.g. $\alpha x(x^2 = 4)$ may denote either $+2$ or -2. If $(E!x)F(x)$, there is only one $\alpha xF(x)$ which is the same as $(\iota x)F(x)$.
Two special relations will be introduced in order to represent relationships expressed in the original text: $A(x, y)$ meaning: 'x occurs in y', and $B(x, y)$ meaning: 'x is the locus of y'. In addition we have: $(x)B(x, y) \rightarrow \rightarrow (z)A(z, \alpha xB(x, y))$.
We can now proceed to a formulation of the theory of proof. The most direct 'means of knowledge' (*pramāna*) is perception. Unfortunately, perception is not always available. Sometimes an object which is not perceptible itself can be *inferred* from a perception. For instance, we may not be able to perceive fire on a distant mountain, but we may perceive smoke; and hence conclude that there is fire because of the smoke.
If a conclusion s can be inferred from a reason h we shall write $V(h, s)$ (with reference to the initial letters of the Sanskrit terms). A proof or inference consists in showing under which conditions $V(h, s)$ holds. Such a proof can be applied if the validity of these conditions can be established by direct perception. Then $V(h, s)$ is valid, and if h is perceived, s may be inferred.
It is said in the first instance that $V(h, s)$ is valid if and only if:
(1) there is an x such that $x \neq s$;
(2) there is a y such that $B(y, x)$, where for x the condition (1) holds;
(3) $\neg A(h, y)$, where for y the condition (2) holds.
In other words a first definition of inference can be written as follows:

$$V(h, s) \leftrightarrow \neg A(h, \alpha yB(y, \alpha x(x \neq s))) \quad \text{(Def. I).}$$

Another definition interchanges the order of the conditions (2) and (3) and can accordingly be written as follows:

$$V(h, s) \leftrightarrow \neg A(h, \alpha y(y \neq \alpha xB(x, s))) \quad \text{(Def. II).}$$

These formulas have been constructed in such a way that there is an isomorphism between the formulas and the Sanskrit expressions in the original. The possibility of this construction is partly due to the use of the α-terminology. Another isomorphism may be established between the formulas and the following figures.
In terms of these schemes $V(h, s)$ is proved whenever it is possibile to establish the validity of the three steps which lead from s to h in the direction indicated by the arrows. The validity in each of these three cases can be established from direct perception.
This approach is not very different from the interpretation of a part of

mathematics as a set of inferences of the form: 'if the axioms A_1, A_2, ..., A_n are valid, the theorem T_k is valid', etc. In both cases the inference is formulated in all generality, whether the premiss is valid or not. In both cases the validity of the inference implies that the conclusion holds whenever the initial conditions or axioms hold.

Def. I.

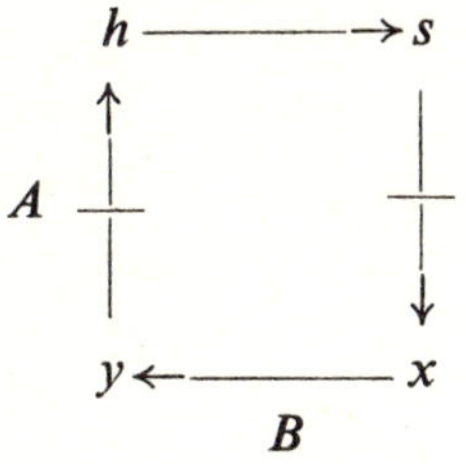

Def. II.

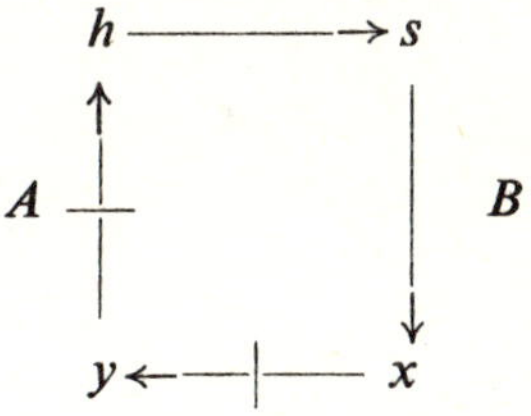

EXAMPLE

V (smoke, fire) is valid:

Def. I.
$ax(x \neq s)$: absence of fire.
$ayB(y$, absence of fire$)$: lake.
$\neg$ *A* (smoke, lake).

Def. II.
$axB(x, s)$: kitchen.
$ay(y \neq$ kitchen$)$: lake
$\neg$ *A* (smoke, lake).

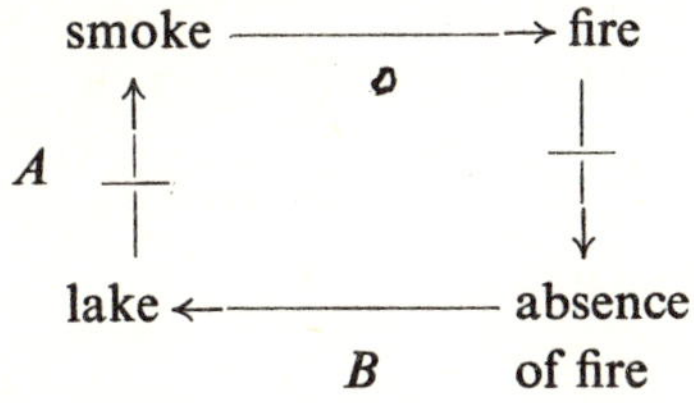

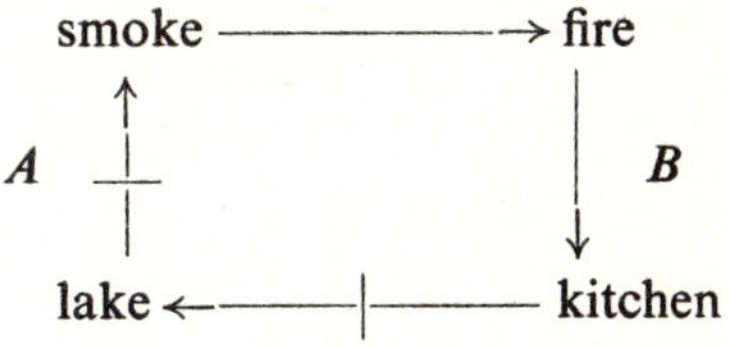

COUNTEREXAMPLE

V (fire, smoke) is invalid:

Def. I.
$ax(x \neq s)$: non-smoke.
$ayB(y$, non-smoke$)$:
red-hot iron bar.
A (fire, red-hot iron bar).

Def. II.
$axB(x, s)$: smoky place.
$ay(y \neq$ smoky place$)$:
red-hot iron bar.
A (fire, red-hot iron bar).

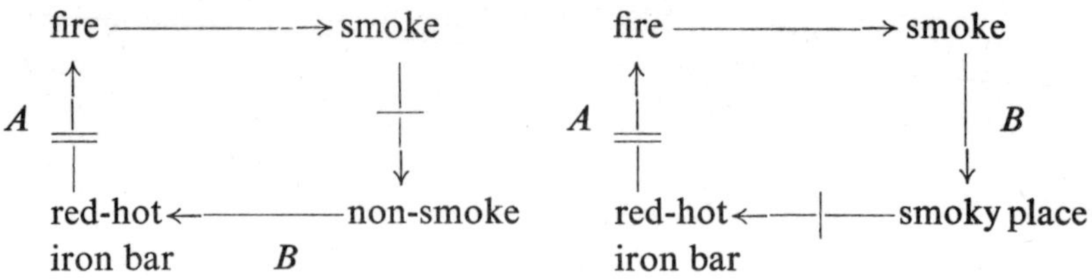

After applying these and other similar definitions to many cases and submitting them to various tests, some definitions are accepted whilst others are rejected. The two definitions mentioned here are rejected for two main reasons, which will be shortly referred to.
(A) If $(y)B(y, x)$ then: $(z)A(z, \alpha yB(y, x))$ or: $(z)A(z, \alpha yB(y, \alpha x(x \neq s)))$, which contradicts definition I. Similarly, if $(x)B(x, s)$ then: $\neg\,(\mathrm{E}y)(y \neq \alpha xB(x, s))$, which prevents the application of definition II.
Such so-called unnegatable or omnipresent terms, defined by $(x)B(x, s)$, are actually available: for instance 'knowable', which may also occur as a conclusion of a proof, for instance in the inference: V (namable, knowable). Hence the definition should account for the validity of such an inference, although it is unable to do so.
(B) There are other cases where the definition does not enable us to prove a conclusion which seems to be intuitively acceptable. As the example produced by the Indian logicians involves abstruse Nyāya categories, a modern example may illustrate the difficulty. Consider again the valid inference V (smoke, fire):
$\alpha xB(x, \text{fire})$: kitchen.
$\alpha y(y \neq \text{kitchen})$: my mind.
A (smoke, my mind).
Here we have played a kind of trick: while 'kitchen' is undoubtedly different from 'my mind', 'smoke' must occur in 'my mind' whenever I think of smoke. Hence the inference seems to be shown to be invalid, though it should be valid. The difficulty lies in the kind of occurrence of 'smoke' in 'my mind'. What is evidently needed is a further precision of the occurrence relation A: the manner in which 'smoke' occurs in 'my mind' when I think about it is different from the manner in which smoke generally occurs, as exemplified by its occurrence (or non-occurrence) in the kitchen. Now the general place where something occurs whenever it occurs 'properly' is called its *residence* and will be denoted by p. The

additional condition, which should hold in order that the definition be valid, is that h occurs through A in the same manner in which it occurs in p. The different kinds of A can now be distinguished by means of bracketed subscripts, such as: $A_{(x)}$, $A_{(y)}$, ... Then the A in definition I (and analogously in definition II), should be specified as follows:

$$azA_{(z)}(h, p) = azA_{(z)}(h, ayB(y, ax(x \neq s)))$$

or:

$$V(h, s) \leftrightarrow \neg A_{(azA_{(z)}(h,p))}(h, ayB(y, ax(x \neq s))).$$

If this is applied to the case of occurrence of 'smoke' in 'my mind', it is evident that: $azA_{(z)}$ (smoke, my mind) $\neq$ $azA_{(z)}$ (smoke, p).
Hence: $\neg A_{(azA_{(z)}(\text{smoke}, p))}$ (smoke, $ay(y \neq axB(x$, fire))), which establishes the validity of V (smoke, fire).

There are several other insertions to the original definitions, enabling them to meet various tests. One source of difficulties is the lack of quantification, which in the above was partly expressed by the absence of quantifiers and partly by the ambiguity inherent in the expression $axF(x)$. Several insertions consist therefore of gradual quantifications [1]). On the whole many definitions were studied and compared on their respective merits. Some were referred to by special names, such as 'the tiger', 'the lion' – the authors being nicknamed the Tiger-cub and the Lion-cub. The objection of the unnegatables applies to several definitions and does not seem to have been challenged itself: the final definition, which is accepted after all the others have been convincingly refuted, does not make use of negative expressions.
The study of Navya-nyāya logic is still in its infancy. Of the huge mass of manuscript material only a fragment has been published. Even of the considerable amount of published material only a small part is read. Yet the study of this logic is indispensable for an understanding of the later phases of Indian philosophy.
To Western logicians Indian logic may be interesting because it developed into a formal logic without being influenced by Western logic and starting from an entirely different background. In studying the problem of the

[1]) For instances cf. the present author's 'Means of formalisation in Indian and Western logic', Proceedings of the XIIth International Congress of Philosophy (Venice 1958).

universality of logical principles, or the question of the relation between logic and language, it is a great advantage to be able to look beyond the horizon of Western formal logic to the formal logic of India: 'denn sie', says Bocheński [1]), '– und sie allein – bietet dem Historiker eine Möglichkeit von höchster Bedeutung, nämlich die des Vergleichs'.

University of Pennsylvania, Philadelphia

1) Op. cit., 486.

PATRICK SUPPES

A COMPARISON OF THE MEANING AND USES OF MODELS IN MATHEMATICS AND THE EMPIRICAL SCIENCES

I. MEANING

Consider the following quotations:
'A possible realization in which all valid sentences of a theory T are satisfied is called a model of T' (Tarski [1953, p. 11]).
'In the fields of spectroscopy and atomic structure, similar departures from classical physics took place. There had been accumulated an overwhelming mass of evidence showing the atom to consist of a heavy, positively charged nucleus surrounded by negative, particle-like electrons. According to Coulomb's law of attraction between electric charges, such a system will collapse at once unless the electrons revolve about the nucleus. But a revolving charge will, by virtue of its acceleration, emit radiation. A mechanism for the emission of light is thereby at once provided.
'However, this mechanism is completely at odds with experimental data. The two major difficulties are easily seen. First, the atom in which the electrons revolve continually should emit light all the time. Experimentally, however, the atom radiates only when it is in a special, 'excited' condition. Second, it is impossible by means of this model to account for the occurrence of spectral lines of a single frequency (more correctly, of a narrow range of frequencies). The radiating electron of our model would lose energy; as a result it would no longer be able to maintain itself at the initial distance from the nucleus, but fall in toward the attracting center, changing its frequency of revolution as it falls. Its orbit would be a spiral ending in the nucleus. By electrodynamic theory, the frequency of the radiation emitted by a revolving charge is the same as the frequency of revolution, and since the latter changes, the former should also change. Thus our model is incapable of explaining the sharpness of spectral lines' (Lindsay and Margenau [1936, pp. 390–91]).
'The author [Gibbs] considers his task not as one of establishing physical theories directly, but as one of constructing statistic-mechanical models

which have some analogies in thermodynamics and some other parts of physics; hence he does not hesitate to introduce some very special hypotheses of a statistical character' (Khinchin [1949, p. 4]).

'Thus, the model of rational choice as built up from pair-wise comparisons does not seem to suit well the case of rational behavior in the described game situation' (Arrow [1951, p. 21]).

'In constructing the model we shall assume that each variable is some kind of average or aggregate for members of the group. For example, D might be measured by locating the opinions of group members on a scale, attaching numbers to scale positions and calculating the standard deviation of the members' opinions in terms of these numbers. Even the intervening variables, although not directly measured, can be thought of as averages of the values for individual members' (Simon [1957, p. 116]).

'This work on mathematical models for learning has not attempted to formalize any particular theoretical system of behavior; yet the influences of Guthrie and Hull are most noticeable. Compared with the older attempts at mathematical theorizing, the recent work has been more concerned with detailed analyses of data relevant to the models and with the design of experiments for directly testing quantitative predictions of the models' (Bush and Estes [1959, p. 3]).

'I shall describe... various criteria used in adopting a mathematical model of an observed stochastic process... For example, consider the number of cars that have passed a given point by time t. The first hypothesis is a typical mathematical hypothesis, suggested by the facts and serving to simplify the mathematics. The hypothesis is that the stochastic process of the model has independent increments... The next hypothesis, that of stationary increments, states that, if $s < t$, the distribution of $x(t) - x(s)$ depends only on the time interval length $t - s$. This hypothesis means that we cannot let time run through both slack and rush hours. Traffic intensity must be constant.

'The next hypothesis is that events occur one at a time. This hypothesis is at least natural to a mathematician. Because of limited precision in measurements it means nothing to an observer... The next hypothesis is of a more quantitative kind, which also is natural to anyone who has seen Taylor's theorem. It is that the probability that at least one car should pass in a time interval of length h should be $ch + o(h)$' (Doob [1960, p. 27]).

The first of these quotations is taken from a book on mathematical logic, the next two from books on physics, the following three from works on the social sciences, and the last one from an article on mathematical statistics. Additional uses of the word 'model' could easily be collected in another batch of quotations. One of the more prominent senses of the word missing in the above quotations is the very common use in physics and engineering of 'model' to mean an actual physical model as, for example, in the phrases 'model airplane' and 'model ship'.

It may well be thought that it is impossible to put under one concept the several uses of the word 'model' exhibited by these quotations. It would, I think, be too much to claim that the word 'model' is being used in exactly the same sense in all of them. The quotation from Doob exhibits one very common tendency, namely, to confuse or to amalgamate what logicians would call the model and the theory of the model. It is very widespread practice in mathematical statistics and in the behavioral sciences to use the word 'model' to mean the set of quantitative assumptions of the theory, that is, the set of sentences which in a precise treatment would be taken as axioms, or, if they are themselves not adequately exact, would constitute the intuitive basis for formulating a set of axioms. In this usage a model is a linguistic entity and is to be contrasted with the usage characterized by the definition from Tarski, according to which a model is a non-linguistic entity in which a theory is satisfied.

There is also a certain technical usage in econometrics of the word 'model' that needs to be noted. In the sense of the econometricians a model is a class of models in the sense of logicians, and what logicians call a model is called by econometricians a *structure*.

It does not seem to me that these are serious difficulties. I claim that the concept of model in the sense of Tarski may be used without distortion and as a fundamental concept in all of the disciplines from which the above quotations are drawn. In this sense I would assert that the meaning of the concept of model is the same in mathematics and the empirical sciences. The difference to be found in these disciplines is to be found in their use of the concept. In drawing this comparison between constancy of meaning and difference of use, the sometimes difficult semantic question of how one is to explain the meaning of a concept without referring to its use does not actually arise. When I speak of the meaning of the concept of a model I shall always be speaking in well-defined technical contexts and what I shall be claiming is that, given this technical meaning of the concept of model, mathematicians ask a certain kind of question about models and empirical scientists tend to ask another kind of question.

Perhaps it will be useful to defend this thesis about the concept of model by analyzing uses of the word in the above quotations. As already indicated, the quotation from Tarski represents a standard definition of 'model' in mathematical logic. Our references to models in pure mathematics will, in fact, be taken to refer to mathematical logic, that branch of pure mathematics explicitly concerned with the theory of models. The technical notion of possible realization used in Tarski's definition need not be expounded here. Roughly speaking, a possible realization of a theory is a set-theoretical entity of the appropriate logical type. For example, a possible realization of the theory of groups is any ordered couple whose first member is a set and whose second member is a binary operation on this set. The intuitive idea of a model as a possible realization in which a theory is satisfied is too familiar in the literature of mathematical logic to need recasting. The important distinction that we shall need is that a theory is a linguistic entity consisting of a set of sentences, and models are non-linguistic entities in which the theory is satisfied (an exact definition of theories is also not necessary for our uses here).

I would take it that the use of the notion of models in the quotation from Lindsay and Margenau could be recast in these terms in the following manner. The orbital theory of the atom is formulated as a theory. The question then arises, does a possible realization of this theory in terms of entities defined in close connection with experiments actually constitute a model of the theory, or, put another way which is perhaps simpler, do models of an orbital theory correspond well to data obtained from physical experiments with atomic phenomena? It is true that many physicists want to think of a model of the orbital theory of the atom as being more than a certain kind of set-theoretical entity. They envisage it as a very concrete physical thing built on the analogy of the solar system. I think it is important to point out that there is no real incompatibility in these two viewpoints. To define formally a model as a set-theoretical entity which is a certain kind of ordered tuple consisting of a set of objects and relations and operations on these objects is not to rule out the physical model of the kind which is appealing to physicists, for the physical model may be simply taken to define the set of objects in the set-theoretical model. Because of the importance of this point it may be well to illustrate it in somewhat greater detail. We may axiomatize classical particle mechanics in terms of the five primitive notions of a set P of particles, an interval T of real numbers corresponding to elapsed times, a position function s defined on the Cartesian product of the set of particles and the time interval, a mass function m defined on the set of particles, and a force function f defined on the Cartesian product of the set of particles, the time interval and the set of positive integers (the set of positive integers enters into the definition of the force function simply in order to provide a method of naming the forces). A possible realization of the axioms of classical particle mechanics, that is, of the theory of classical particle mechanics, is then an ordered quintuple $\mathscr{P} = < P, T, s, m, f >$. A model of classical particle mechanics is such an ordered quintuple. It is simple enough to see how an actual physical model in the physicist's sense of classical particle mechanics is related to this set-theoretical sense of models. We simply can take the set of particles to be in the case of the solar system the set of planetary bodies. Another slightly more abstract possibility is to take the set of particles to be the set of centers of mass of the planetary bodies. This generally exemplifies the situation. The abstract set-theoretical model of a theory will have among its parts a basic set which will consist of the objects ordinarily thought to constitute the physical model (for a discussion of the axiomatic foundations of classical particle mechanics in greater detail along the lines just suggested see Suppes [1957, Ch. 12]).

In the preceding paragraph we have used the phrases, 'set-theoretical model' and 'physical model'. There would seem to be no use in arguing about which use of the word 'model' is primary or more appropriate in the empirical sciences. My own contention in this paper is that the set-theoretical usage is the more fundamental. The highly physically minded or empirically minded scientists who may disagree with this thesis and believe that the notion of a physical model is the more important thing in a given branch of empirical science may still agree with the systematic remarks I am making.

An historical illustration of this point is Kelvin's and Maxwell's efforts to find a mechanical model of electromagnetic phenomena. Without doubt they both thought of possible models in a literal physical sense, but it is not difficult to recast their published memoirs on this topic into a search for set-theoretical models of the theory of continuum mechanics which will account for observed electromagnetic phenomena. Moreover, it is really the formal part of their memoirs which has had permanent value. Ultimately it is the mathematical theory of Maxwell which has proved important, not the physical image of an ether behaving like an elastic solid.

The third quotation is from Khinchin's book on statistical mechanics, and the phrase, 'the author', refers to Gibbs, whom Khinchin is discussing at this point. The use of the word 'model' in the quotation of Khinchin is particularly sympathetic to the set-theoretical viewpoint, for Khinchin is claiming that in his work on the foundations of statistical mechanics Gibbs was not concerned to appeal directly to physical reality or to establish true physical theories but rather to construct models or theories having partial analogies to the complicated empirical facts of thermodynamics and other branches of physics. Again in this quotation we have as in the case of Doob, perhaps even more directly, the tendency toward a confusion of the logical type of theories and models, but again this does not create a difficulty. Anyone who has examined Gibbs's work or Khinchin's will readily admit the ease and directness of formulating their work in such a fashion as to admit explicitly and exactly the distinction between theories and models made in mathematical logic. The abstractness of Gibbs's work in statistical mechanics furnishes a particularly good example for applying the exact notion of model used by logicians, for there is not a direct and immediate tendency to think of Gibbs's statistical mechanical theories as being the theories of the one physical universe.

I think the following observation is empirically sound concerning the use of the word 'model' in physics. In old and established branches of physics which correspond well with the empirical phenomena they attempt to explain, there is only a slight tendency ever to use the word 'model'. The language of theory, experiment and common sense is blended into one realistic whole. Sentences of the theory are asserted as if they are the one way of describing the universe. Experimental results are described as if there were but one obvious language for describing them. Notions of common sense, refined perhaps here and there, are taken to be appropriately homogeneous with the physical theory. On the other hand, in those branches of physics which give as yet an inadequate account of the detailed physical phenomena with which they are concerned there is a much more frequent use of the word 'model'. Connotation of the use of the word is that the model is like a model of an airplane or ship. It simplifies drastically the true physical phenomena and only gives account of certain major or important aspects of it. Again, in such uses of the word 'model' it is to be emphasized that there is a constant interplay between the model as a physical or non-linguistic object and the model as a theory. The quotation from Arrow which follows the one from Khinchin exemplifies in the social sciences this latter tendency in physics. Arrow, I would say, refers to the *model* of rational choice because the theory he has in mind does not give a very adequate description of the phenomena with which it is concerned but only provides a highly simplified schema. The same remarks apply fairly well to the quotation from Simon. In Simon we have an additional phenomenon exemplified which is very common in the social and behavioral sciences. A certain theory is stated in broad and general terms. Some qualitative experiments to test this theory are performed. Because of the success of these experiments scientists interested in more quantitative and exact theories then turn to what is called 'the construction of a model' for the original theory. In the language of logicians, it would be more appropriate to say that rather than constructing a model they are interested in constructing a quantitative theory to match the intuitive ideas of the original theory.

In the quotation from Bush and Estes and the one from Doob there is introduced an important line of thought which is, in fact, very closely connected with the concept of model as used by logicians. I am thinking here of the notion of model in mathematical statistics, the extensive literature on estimating parameters in models and testing hypotheses about them. In a statistical discussion of the estimation of the parameters of a model it is usually a trivial task to convert the discussion into one where the usage of terms is in complete agreement with that of logicians. The detailed consideration of statistical questions almost requires the consideration of models as mathematical or set-theoretical rather than simple physical entities. The question, 'How well does the model fit the data?' is a natural one for statisticians and behavioral scientists. Only recently has it begun to be so for physicists, and it is still true that much of the experimental literature in physics is reported in terms of a rather medieval brand of statistics.

It may be felt by some readers that the main difficulty with the thesis being advanced in this paper is the lack of substantive examples in the empirical sciences. Such a reader would willingly admit that there are numerous examples of exactly formulated theories in pure mathematics and thereby an exact basis is laid for precisely defining the models in which these theories are satisfied. But it might be held the situation is far different in any branch of empirical sciences. The formulation of theory goes hand in hand with the development of new experiments and new experimental techniques. It is the practice of empirical scientists, so it might be claimed, not to formulate theories in exact fashion but only to give them sufficient conceptual definiteness to make their connections with current experiments sufficiently clear to other specialists in the field. He who seeks an exact characterization of the theory and thus of models in such branches of science as non-vertebrate anatomy, organic chemistry or nuclear physics is indeed barking up the wrong tree. In various papers and books I have attempted to provide some evidence against this view. In the final chapter of my *Introduction to Logic* I have formulated axiomatically a theory of measurement and a version of classical particle mechanics which satisfy, I believe, the standards of exactness and clarity customary in the axiomatic formulation of theories in pure mathematics. In Estes and Suppes [1960] such a formulation is attempted for a branch of mathematical learning theory. In Rubin and Suppes [1954] an exact formulation of relativistic mechanics is considered and in Suppes [1959] such a formulation of relativistic kinematics is given. These references are admittedly egocentric; it is also pertinent to refer to the work of Woodger [1937], Hermes [1938], Adams [1959], Debreu [1959], Noll [1959] and many others. Although it is not possible to pinpoint a reference to every branch of empirical science which will provide an exact formulation of the fundamental theory of the discipline, sufficient examples do now exist to make the point that there is no systematic difference between the axiomatic formulation of theories in well-developed branches of empirical science and in branches of pure mathematics.

By remarks made from a number of different directions I have tried to argue that the concept of model used by mathematical logicians is the basic and fundamental concept of model needed for an exact statement of any branch of empirical science. To agree with this thesis it is not necessary to rule out or to deplore variant uses or variant concepts of model now abroad in the empirical sciences. As has been indicated, I am myself prepared to admit the significance and practical importance of the notion of physical model current in much discussion in physics and engineering. What I have tried to claim is that in the exact statement of the theory or in the exact analysis of data the notion of model in the sense of logicians provides the appropriate intellectual tool for making the analysis both precise and clear.

II. USES

The uses of models in pure mathematics are too well-known to call for review here. The search in every branch of mathematics for representation theorems is most happily characterized in terms of models. To establish a representation theorem for a theory is to prove that there is a class of models of the theory such that every model of the theory is isomorphic to some member of this class. Examples now classical of such representation theorems are Cayley's theorem that every group is isomorphic to a group of transformations and Stone's theorem that every Boolean algebra is isomorphic to a field of sets. Many important problems in mathematical logic are formulated in terms of classes of models. For a statement of many interesting results and problems readers are referred to Tarski [1954].

When a branch of empirical science is stated in exact form, that is, when the theory is axiomatized within a standard set-theoretical framework, the familiar questions raised about models of the theory in pure mathematics may also be raised for models of the precisely formulated empirical theory. On occasion such applications have philosophical significance. Many of the discussions of reductionism in the philosophy of science may best be formulated as a series of problems using the notion of a representation theorem. For example, the thesis that biology may be reduced to physics would be in many people's minds appropriately established if one could show that for any model of a biological theory it was possible to construct an isomorphic model within physical theory. The diffuse character of much biological theory makes any present attempt to realize such a program rather hopeless. An exact result of this character can be established for one branch of physics in relation to another. An instance of this is Adams's [1959] result that for a suitable characterization of rigid body mechanics every model of rigid body mechanics is isomorphic to a model defined within simple particle mechanics. But I do not want to give the impression that the application of models in the empirical sciences is mainly restricted to problems which interest philosophers of science. The attempt to characterize exactly models of an empirical theory almost inevitably yields a more precise and clearer understanding of the exact character of the theory. The emptiness and shallowness of many classical theories in the social sciences is well brought out by the attempt to formulate in any exact fashion what constitutes a model of the theory. The kind of theory which mainly consists of insightful remarks and heuristic slogans will not be amenable to this treatment. The effort to make it exact will at the same time reveal the weakness of the theory.

An important use of models in the empirical sciences is in the construction of Gedanken experiments. A Gedanken experiment is given precision and clarity by characterizing a model of the theory which realizes it. A standard and important method for arguing against the general plausibility of a theory consists of extending it to a new domain by constructing a model of the theory in that domain. This aspect of the use of models need not however be restricted to Gedanken experiments. A large number of experiments in psychology are designed with precisely this purpose in mind, that is, the extension of some theory to a new domain, and the experimenter's expectation is that the results in this domain will not be those predicted by the theory.

It is my own opinion that a more exact use of the theory of models in the discussion of Gedanken experiments would often be of value in various branches of empirical science. A typical example would be the many discussions centering around Mach's proposed definition of the mass of bodies in terms of their mutually induced accelerations. Because of its presumed simplicity and beauty this definition is frequently cited. Yet from a mathematical standpoint, and on any exact theory of models of the theory of mechanics, it is not a proper definition at all. For a very wide class of axiomatizations of classical particle mechanics it may be proved by Padoa's principle that a proper definition of mass is not possible. Moreover, if the number of interacting bodies is greater than seven a knowledge of the mutually induced acceleration of the particles is not sufficient for unique determination of the ratios of the masses of the particles. The fundamental weakness of Mach's proposal is that he did not seem to realize a definition in the theory cannot be given for a single model, but must be appropriate for every model of the theory in order to be acceptable in the standard sense.

Another significant use of models, perhaps peculiar to the empirical sciences, is in the analysis of the relation between theory and experimental data. The importance of models in mathematical statistics has already been mentioned. The homogeneity of the concept of model used in that discipline with that adopted by logicians has been remarked upon. The striking thing about the statistical analysis of data is that it is shot through and through with the kind of comparison of models that does not ordinarily arise in pure mathematics. Generally speaking, in some particular branch of pure mathematics the comparison of models involves comparison of two models of the same logical type. The representation theorems mentioned earlier provide good examples. Even in the case of embedding theorems, which establish that models of one sort may be extended in a definite manner to models of another sort, the logical type of the two models is very similar. The situation is often radically different in the comparison of theory and experiment. On the one hand, we may have a rather elaborate set-theoretical model of the theory which contains continuous functions or infinite sequences, and, on the other hand, we have highly finitistic set-theoretical models of the data. It is perhaps necessary to explain what I mean by 'models of the data'. The maddeningly diverse and complex experience which constitutes an experiment is not the entity which is directly compared with a model of a theory. Drastic assumptions of all sorts are made in reducing the experimental experience, as I shall term it, to a simple entity ready for comparison with a model of the theory.

Perhaps it would be well to conclude with an example illustrating these general remarks about models of the data. I shall consider the theory of linear response models set forth in Estes and Suppes [1959]. For simplicity, let us assume that on every trial the organism can make exactly one of two responses, A_1 or A_2, and after each response it receives a reinforcement, E_1 or E_2, of one of the two possible responses. A learning parameter Θ, which is a real number such that $0 < \Theta \leq 1$, describes the rate of learning, in a manner to be made definite in a moment. A possible realization of the theory is an ordered triple $\mathscr{X} = < X, P, \Theta >$ of the following sort. X is the set of all sequences of ordered pairs such that the first member of each pair is an element of some set A and the second member an element of some set B, where A and B each have two elements. Intuitively, the set A represents the two possible responses and the set B the two possible reinforcements. P is a probability measure on the Borel field of cylinder sets of X, and Θ is a real number as already described. (Actually there is a certain arbitrariness in the characterization of possible realizations of theories whose models have a rather complicated set-theoretical structure, but this is a technical matter into which we shall not enter here.) To define the models of the theory, we need a certain amount of notation. Let $A_{i,n}$ be the event of response A_i on trial n, $E_{j,n}$ the event of reinforcement E_j on trial n, where $i, j = 1, 2$, and for x in X let x_n be the equivalence class of all sequences in X which are identical with x through trial n. A possible realization of the linear response theory is then a model of the theory if the following two axioms are satisfied in the realization:

Axiom 1. *If* $P(E_{i,n}A_{i',n}x_{n-1}) > 0$ *then*

$$P(A_{i,n+1} \mid E_{i,n}A_{i',n}x_{n-1}) = (1 - \Theta)P(A_{i,n} \mid x_{n+1}) + \Theta.$$

Axiom 2. *If* $P(E_{j,n}A_{i',n}x_{n-1}) > 0$ *and* $i \neq j$ *then*

$$P(A_{i,n+1} \mid E_{j,n}A_{i',n}x_{n-1}) = (1 - \Theta)P(A_{i,n} \mid x_{n-1}).$$

As is clear from the two axioms, this linear response theory is intuitively very simple. The first axiom just says that when a response is reinforced the probability of making that response on the next trial is increased by a simple linear transformation. And the second axiom says that if some other response is reinforced, the probability of making the response is decreased by a second linear transformation. In spite of the simplicity of this theory it gives a reasonably good account of a number of experiments, and from a mathematical standpoint it is by no means trivial to characterize asymptotic properties of its models.

The point of concern here, however, is to relate models of this theory to models of the data. Again for simplicity, let us consider the case of simple noncontingent reinforcement. On every trial the probability of an E_1 reinforcement, independent of any preceding events, is π. The experimenter decides on an experiment of, say, 400 trials for each subject, and he uses a table of random numbers to construct for each subject a finite reinforcement sequence of 400 trials. The experimental apparatus might be described as follows.

The subject sat at a table of standard height. Mounted vertically on the table top was a 125 cm. wide by 75 cm. high black panel placed 50 cm. from the end of the table. The experimenter sat behind the panel, out of view of the subject. The apparatus, as viewed by the subject, consisted of two silent operating keys mounted 20 cm. apart on the table top and 30 cm. from the end of the table; upon the panel, three milk-glass panel lights were mounted. One of these lights, which served as the signal for the subject to respond, was centered between the keys at a height of 42 cm. from the table top. Each of the two remaining lights, the reinforcing signals, was at a height of 28 cm. directly above one of the keys. On all trials the signal light was lighted for 3.5 sec.; the time between successive signal exposures was 10 sec. The reinforcing light followed the cessation of the signal light by 1.5 sec. and remained on for 2 sec.

The model of the data incorporates very little of this description. On each trial the experimenter records the response made and the reinforcement given. Expressions on the subject's face, the movement of his limbs, and in the present experiment even how long he takes to make the choice of which key to punch, are ignored and not recorded. Even though it is clear exactly what the experimenter records, the notion of a possible realization of the data is not unambiguously clear. As part of the realization it is clear we must have a finite set D consisting of all possible finite sequences of length 400 where, as previously, the terms of the sequences are ordered couples, the first member of each couple being drawn from some pair set A and the second member from some pair set B. If a possible realization consisted of just such a set D, then any realization would also be a model of the data. But it seems natural to include in the realization a probability measure P on the set of all subsets of D, for by this means we may impose upon models of the data the experimental schedule of reinforcement. In these terms, a possible realization of the data is an ordered couple $\mathscr{D} = < D, P >$ and, for the case of noncontingent reinforcement a realization is a model if and only if the probability measure P has the property of being a Bernoulli distribution with parameter π on the second members of the terms of the finite sequences in D, i.e. if and only if for every n from 1 to 400, $P(E_{1,n} \mid x_{n-1}) = \pi$ when $P(x_n) > 0$.

Unfortunately, there are several respects in which this characterization of models of the data may be regarded as unsatisfactory. The main point is that the models are still too far removed even from a highly schematized version of the experiment. No account has been taken of the standard practice of randomization of response A_1 as the left key for one subject and the right key for another. Secondly, a model of the data, as defined above, contains 2^{400} possible response sequences. An experiment that uses 30 or 40 subjects yields but a small sample of these possibilities. A formal description of this sample is easily given, and it is easily argued that the 'true' model of the data is this actual sample, not the much larger model just defined. Involved here is the formal relation between the three entities labeled by statisticians the 'sample', the 'population', and the 'sample space'. A third difficulty is connected with the probability measure that I have included as part of the model of the data. It is certainly correct to point out that a model of the data is hardly appropriately experimental if there is no indication given of how the probability distribution on reinforcements is produced.

It is not possible in this paper to enter into a discussion of these criticisms or the possible formal modifications in models of the data which might be made to meet them. My own conviction is that the set-theoretical concept of model is a useful tool for bringing formal order into the theory of experimental design and analysis of data. The central point for me is the much greater possibility than is ordinarily realized of developing an adequately detailed formal theory of these matters.

Stanford University, Stanford, Cal.

REFERENCES

Adams, E. W. The foundations of rigid body mechanics and the derivation of its laws from those of particle mechanics. In L. Henkin, P. Suppes and A. Tarski (Eds.), *The Axiomatic Method.* Amsterdam: North-Holland Publishing Co., 1959.

Arrow, K. J. *Social Choice and Individual Values.* (Cowles Commission Monograph 12) New York: Wiley, 1951.

Bush, R. R., and Estes, W. K. (Eds.). Introduction in *Studies in Mathematical Learning Theory.* Stanford, Calif.: Stanford Univ. Press, 1959.

Debreu, G. *Theory of Value.* (Cowles Foundation Monograph 17) New York: Wiley, 1959.

Doob, J. L. Some problems concerning the consistency of mathematical models. In R. E. Machol (Ed.), *Information and Decision Processes.* New York: McGraw-Hill, 1960.

Estes, W. K., and Suppes, P. Foundations of linear models. Ch. 8 in Bush, R. R., and Estes, W. K. (Eds.), *Studies in Mathematical Learning Theory.* Stanford, Calif.: Stanford Univ. Press, 1959.

Hermes, H. Eine Axiomatisierung der allgemeinen Mechanik, Forschungen zur Logik und zur Grundlegung der exakten Wissenschaften, new series, 3 (1938), S. Hirzel, Leipzig, 48 pp.

Khinchin, A. I. *Statistical Mechanics.* New York: Dover, 1949.

Lindsay, R. B., and Margenau, H. *Foundations of Physics.* New York: Wiley, 1936.

Noll, W. The foundations of classical mechanics in the light of recent advances in continuum mechanics. In L. Henkin, P. Suppes and A. Tarski (Eds.), *The Axiomatic Method.* Amsterdam: North-Holland Publishing Co., 1959.

Rubin, H., and P. Suppes. Transformations of systems of relativistic particle mechanics. *Pacific Journal of Mathematics,* 4 (1954), pp. 563–601.

Simon, H. A. *Models of Man.* New York: Wiley, 1957.

Suppes, P. *Introduction to Logic.* Princeton: Van Nostrand, 1957.

Suppes, P. Axioms for relativistic kinematics with or without parity. In L. Henkin, P. Suppes and A. Tarski (Eds.), *The Axiomatic Method.* Amsterdam: North-Holland Publishing Co., 1959.

Tarski, A. A general method in proofs of undecidability. Ch. I in A. Tarski, A. Mostowski, and R. M. Robinson (Eds.), *Undecidable Theories.* Amsterdam: North-Holland Publishing Co., 1953.

Tarski, A. Contributions to the theory of models. *Indagationes Mathematicae,* 16 (1954), pp. 572–588, and 17 (1955), pp. 56–64.

Woodger, J. H. *The Axiomatic Method in Biology.* Cambridge, England: Cambridge Univ. Press, 1957.

J. B. UBBINK

MODEL, DESCRIPTION AND KNOWLEDGE

INTRODUCTION

It is not my intention to speak about the use of models in scientific inquiry or the connections between formal systems, theories and models. The problem I should like to discuss can be loosely stated as follows: Do our scientific books and articles contain descriptions of nature; does the scientist construct a picture or model of nature? Or again: Is our scientific knowledge true if, and only if, it is a model of nature? Is, for instance, Hertz right when he states 'We make for ourselves internal pictures of external objects... When on the basis of our accumulated previous experiences we have succeeded in constructing pictures... we can quickly derive by means of them, as by means of models, the consequences...' – or must we not only believe Dirac when he states: '...the main object of physical science is not the provision of pictures' but go one step farther and say that knowledge of nature has no more to do with a model of nature than religion has to do with idols? Or to put it boldly: does the scientist observe the first part of the third Commandment: 'Thou shalt not make unto thee any graven image, or any likeness of anything that is in the heaven above, or that is in the earth beneath, or that is in the water under the earth...'? And if this should be the case: in what sense can knowledge be true?

The fact that I will not speak about the use of models in scientific inquiry – and in doing so could use the term 'model' in a fairly technical and well-defined sense – renders it advisable to point out beforehand that I will not use the term 'model' (only) for an object that by virtue of its structure or form can represent another object (see Hertz quotation), but that I will extend its meaning in such a way that the term 'model' also covers a sentence or a system of sentences. The essential thing is that a model represents an object or matters of fact in virtue of its structure; so an object is a model (in the narrow and in the wider sense of the term) of matters of fact if, and only if, their structures are isomorphic. In this sense an electrical circuit can be a model of a hydrodynamical circuit, a map a model of a country and so on.

I. THE GREEK INHERITANCE

(1) Sometimes it is a good thing to stick to an opinion which is clearly untenable. It is possible that there may be – to quote Bohr – 'deep truth' in it; moreover it may afterwards become clear that by doing so a discipline is developed which is of the utmost importance though it has to give up the pretension to universal validity.

Now I tend to think that such is the case with the opinion of Parmenides in particular and with the vision of the Greeks in general. Let me be more explicit.

The statements of Parmenides that 'being is one and not many' and that 'being *is*, whereas becoming and change are not real' are – as Parmenides certainly knew – untenable in a superficial sense. But that does not mean that there is no 'deep truth' in this statement: it is true that whoever seeks knowledge seeks unity; we believe that our knowledge of nature has – in a logical sense – to be one: a theory has to possess a logical structure, otherwise it is no theory. (However, the lesson is not – as Parmenides thought – that knowledge should be acquired without recourse to sense-perception which shows us diversity and change, but on the contrary, how to unite this diversity.) Moreover it seems to me to be both a degradation and a misunderstanding of Parmenides' views to cite – as is often done – conservation laws as an example of what he had in mind: Parmenides distinguished 'real' knowledge, which can only be acquired by 'thinking', from 'bare opinions', which are generated by sense-perception; this implies that his statements about the properties of being cannot be brought in line with our statements about the world of physics.

In the second place, it may be remarked that the overestimation of the power of thought in ancient times played an essential part in the minds of those who, for the first time, developed mathematics as a science. That scientific inquiry into nature was frustrated by this overestimation should not stop the modern physicist from being grateful to the Greeks, who, by inventing mathematics and logic, made our scientific inquiry possible. On the other hand the philosopher should never forget that this overestimation of the power of thought set up an ideal of knowledge of nature that cannot be ours; one should not forget that the Greeks were not interested in knowledge of nature in our sense, and that for this reason their knowledge of being as such should not be seen as a model of our physical world. It is particularly necessary to be mindful of this fact if one seeks – as I intend to do now – to find out which difficulties knowledge by perception presented to the old philosophers.

(2) When Heraclitus said: 'one cannot bathe twice in the same river' he was in a sense undoubtedly right: the river on Monday is not identical with the river on Tuesday – it has only a certain similarity to it. In short: Heraclitus means by 'the same' 'identical with' – not 'model of'. His pupil Cratylus went one step farther: he stated that you cannot bathe in the river 'Rhine'; what he meant was – as far as we can say – that one cannot truly say this. And we may draw the conclusion: to use one and the same word (word-identity) for things which do not possess identity of being is, according to Cratylus, inadmissible. Therefore, in fact, it is not admissible to use words for perceptible things at all. One can only point at them.

We may conclude that in early ancient times the character of a description was not clearly understood: it was supposed that the word – as a certain recognizable object – would have to be identical with the thing it designates to be able to speak truly of the thing in question. On this basis we can understand how Gorgias the Sophist could state (1) that we cannot know; (2) that – if it were possible to know – it would not be possible for us to share this knowledge with others by means of language. What should interest us is the conception of knowledge which renders these statements valid. And that is again the notion of identity: knowledge (as a state of the mind) in order to be true has to be identical with its object. Because there is no identity between one's state of mind and the object, one cannot know; because there is no identity between my state of mind and yours we cannot share knowledge and because there is no identity between, e.g., a spoken word 'red' – which itself is certainly not a colour – and the redness of things, descriptions cannot be true.

It is clear that this notion of knowledge, together with the divergence of opinion of different philosophers, destroyed faith in the possibility for man of acquiring objective knowledge. But – so it is said – Socrates and Plato restored this faith. Now I am at the moment not interested in the ethical or the 'deep' philosophical importance of the work of Plato and Socrates – I am interested only in the way in which they transformed the notion of knowledge. In this connection one should remind oneself that Plato was deeply interested in mathematics; the type of knowledge we find in mathematics was for him the only kind of true knowledge. Moreover in mathematics one cannot be bothered by such problems as are mentioned above. Mathematicians – as Plato tells us – make use of drawn figures. What they think about, however, is not these drawings but the figures 'themselves'; their demonstrations concern the square and the diagonal themselves – not any particular drawing of a square and a diagonal; they try to see what can be seen only by the mind's eye – it is only this which can show us the truth (Republic).

On the other hand Plato believes with Heraclitus, Cratylus and the Sophists that the reality we can see with our sense-organs is confused and contradictory. He gives a few more arguments. For instance: for one and the same person we use sometimes the word 'small', sometimes the word 'big'; it is sense-perception which induces us to do these clearly inadmissible things – therefore, he concludes, sense-perception only creates bare opinions, not real knowledge. Or again: I have a cigarette and apply a match to it – I have one thing and add another: what is it that makes them (which subject?) 'two'; or 'how can a subject that is 'one' become 'two'? Therefore, Plato says, what we should try to do is to find out what these things are in themselves. The object of our inquiries has to be the ideal reality, not the confused, intricate and contradictory physical reality.

For the moment, however, I am not concerned with Plato's idealism. Plato was involved in a discussion with his contemporaries. To them he pointed out that a word does not have to be identical with the perceived thing it designates to have any meaning. Mathematicians, for instance, use words in meaningful discourse; they do not, however, speak about things which can be seen by the eye of the body but about things 'themselves', that is to say, about things that can be seen by the eye of the mind. In this sense we may say that, through Plato, the mathematicians rescued knowledge and meaning.

This is by no means all: when Plato says that a man is only a man in virtue of his participation in the idea 'man', he is making an effort to come down to earth again. However, what is meant by this sentence is at this moment only important inasmuch as it enables us to say of a man that he is (or may be) a 'man' irrespective of the fact that there is no identity nor likeness between the spoken word 'man' (which can be heard) and the man as he can be seen. In this sense the net result of Plato's concern was that (i) a word corresponds to the object it designates (it is a label) notwithstanding that (ii) it is not identical with this object. In short: for Plato a word becomes a symbol and not, as for his predecessors, an image. Aristotle – Plato's disciple – was able to make another step forward; he was not a mathematician but a scientist. This implies that he was interested in the way we use words in daily life.

He stated that concepts are formed by means of simple apprehension (without the help of judgement): the 'nous' abstracts from the sensible images the 'quiddity' which is common to them all. At the same time, in agreement with Plato, he also stated that between name and object there is only correspondence. This means that his semantic scheme was a rather simple one: written words are symbols of spoken words, these are symbols of mental experiences, and mental experiences are again symbols of things; so spoken words are symbols of things.

But there is, however, a difficulty here: when I say of a tomato on hand that it is red, there is a one-to-one correspondence between the word and the actual redness. An atomic description of this kind, however, cannot be true, for it does not reveal anything beyond the fact that redness is present: there is neither identity nor similarity between word and fact: 'A sentence is a significant portion of speech, some parts of which have an independent meaning; that is to say, as an utterance, though not as the expression of any positive judgement. Let me explain: The word 'human' has meaning, but does not constitute a proposition, either positive or negative. It is only when other words are added that the whole will form an affirmation or a denial' (*De interpr*. 16b 26–30).

Atomic propositions cannot, but molecular propositions can, precisely in virtue of their complexity, be true (logical truth). Molecular propositions, in virtue of their form, can mirror the form of the (complex) object – they can be models of this object. I quote again (*Metaph. Θ*. 10.1051 b 2–17): 'This (viz. 'being as truth' and 'non-being as falsity') depends, on the side of the objects, on their being combined or separated, so that he who thinks the separated to be separated and the combined to be combined has the truth, while he whose thought is in a state contrary to that of the objects is in error.' And: (*Metaph*. E4 1027 b 18–23): 'But since that which is, in the sense of being true, or is not, in the sense of being false, depends on combination or separation, and truth and falsity together depend on the allocation of a pair of contradictory judgements, for the true judgement affirms where the subject and the predicate really are combined, and denies where they are separated, whereas the false judgement gives the opposite allocation'. In other words: in the proposition the terms are united-and-one; therefore the proposition is true if and only if the objects that are designated by the words are in fact united-and-one. Or: in order to be true, a proposition has to be a model of its object. Summary: a proposition mirrors the being-united-and-one properties of the objects in the being-united-and-one of the words. In this way – that is by way of structure – the correspondence between word and object can serve its function. In a simple description 'This is red' of a red tomato this correspondence cannot be used; in this case only the correspondence which tells us nothing about the object remains.

There is – I think – no need to stipulate that I am not at this moment interested in questions involving realism, conceptualism, nominalism, or whether one has to say that words or propositions concern 'real facts' or 'states of mind'. If we leave these questions out of consideration the preceding, as far as I can see, is in accordance not only with the opinion held by Aristotle and Thomas but also with those of Locke, Hume, Mach, Wittgenstein (see, e.g., Tractatus 2). The question I am here concerned with is whether this traditional point of view is tenable at all; I do not think it is. I think rather that Quine is right when he says: 'My present suggestion is that it is nonsense and the root of much nonsense, to speak of a linguistic component and a factual component in the truth of any individual statement... the unit of empirical science is the whole of science' (From a logical point of view, Two dogmas of empiricism, p. 42).

II. CRITICAL REMARKS

The transition from 'B belongs to (all) A' to 'all A's are B's' indicates a change of background and a re-evaluation of sensorial perception. For Plato and Aristotle the universal proposition 'all men are mortal' should be read as 'man is mortal'. And that means either that this proposition is a proposition about man (a certain matter of fact) in Plato's heaven (and as such does not constitute a problem); or that we have to ask ourselves whether man is necessarily mortal. Now Hume denies firstly that there are individuals in an ontological sense (and this implies that one cannot in truth speak of 'belonging to'); secondly that the togetherness of A's and B's is always accidental (there is no reason for it; the togetherness is not necessary). That is to say: one can never be sure about a continuation of the 'united-and-one'-ness.

To be sure, Hume had to say this; his point of view is implied by the whole set-up; one cannot take interest in our sub-lunary world *and* escape Hume's conclusion without altering the whole idea of linguistic constructions as models of complex facts.

When, for instance, we see that a tomato is red and, by tasting it, find out that it is a ripe one, then the statement 'This tomato is red and ripe' in its union of the words 'red' and 'ripe' serves as a model of the united-and-one-ness of these qualities. But we can never be sure that another red tomato will also taste good, for – according to the traditional view (esp. Thomas) – there is only one legitimate means of asserting that it is ripe: that of tasting it. And it is clear that as long as one only looks, one cannot have a taste sensation. Of course one can have the habit of expecting that a red tomato will taste good – this however is only a habit: one does not know (have a certain taste-sensation) what the taste-sensation will be before one has it. Or to quote Kant on this matter: (*Prolegomena*; Vorwort): ...es ist aber gar nicht einzusehen wie darum, weil Etwas ist, etwas anderes notwendiger Weise auch sein müsse, und wie sich also der Begriff von einer solchen Verknüpfung apriori einführen lasse'.

Before I proceed I will make one remark: the formulation of the universal proposition itself is rather mysterious and contains an illegitimate suggestion. The term 'all' itself is – as Brouwer said – a figment of the brain, an idle fancy, a chimaera; what do we mean by this word in cases such as those mentioned above; what *can* we mean by it? Moreover, this term suggests that we should try to cover – not all, that is impossible – at least 'many' cases, or 'most'; but 'many' is not enough to give the proposition objective validity – it can only induce a habit. To avoid this chimaera and the suggestion I prefer to read this 'universal' proposition as: 'if A then B' (if the tomato is red then it is ripe). This means: I believe that I shall never see a red tomato that is not ripe. (I shall come back to this point later.)

However this may be: the whole set-up implies that universal propositions have a subjective component: they express a belief or habit (and it is not an agreeable thought that our scientific works do not contain 'objective knowledge' but beliefs).

Now I prefer not to make another hypothesis such as – for instance – Mill's principle of uniformity of nature, or Keynes's principle of limited variety; I think that Braithwaite (*Scientific explanation*, p. 259) is right when he says: 'The overwhelming objection to the assimilation of all induction to deduction is that this would require that we should reasonably believe a very general empirical major premiss, the reasonableness of belief in which would have to be justified by another inductive argument.' In other words: such an hypothesis would never be more than an *ad hoc* hypothesis. Therefore it seems to me to be more adequate, that is to say to be more in accordance with the spirit of scientific inquiry, to acknowledge the fact that the traditional theory of observation (simple apprehension) and description is not adequate as a basis for an account of the existence and the role of universal propositions; to acknowledge moreover the fact that we cannot abolish such propositions and have to see them as 'objective' or 'subjective' in exactly the same sense as our other descriptions. This means that we have to try to abolish or to refashion the traditional theory.

At this point a side remark may be allowed. Let me assume that you did not only come to Holland for scientific reasons but also because you are interested in the Dutch way of life and in the Dutch. Now – as a tourist – you know that 'Dutchmen wear wooden shoes'. You see however that I am not wearing wooden shoes; therefore you think that I am not a Dutchman (Mr Apostel, for instance – seeing that I was not wearing wooden shoes – yesterday addressed me in English). But I, in my turn, show you my passport and you can see that I am a Dutchman. Now I think that you are – as a good scientist of the Humean tradition – very happy because it is demonstrated again that a universal proposition is not valid. But it is not as simple as that. For if you are a real tourist you will certainly say: that demonstrates only that you are not a real (or typical) Dutchman, for 'Dutchmen wear wooden shoes' whereas you do not. On account of this you will lose all interest in me because you came to Holland to meet real Dutchmen.

Let me now make an identification: we will call the tourist Aristotle, the scientist Hume; let me further say that Aristotle uses his knowledge ('Dutchmen wear wooden shoes') as a norm. Then it is clear that the issue between Aristotle and Hume cannot be decided *in abstracto* and in general: all we can say is that Hume is trying to falsify the statement and – therefore – does not use it as a rule; whereas Aristotle uses the statement as a rule, and therefore does not test it. Now Hume assumes that empirical statements should be tested – but *why* should this be so? There certainly is a very good reason: As it is based on phenomena, it must be possible to repudiate it by phenomena. Aristotle however can reply: if, and only if, Mr Ubbink is a Dutchman, the fact that he does not wear wooden shoes would falsify my opinion about Dutchmen – as, however, he is no Dutchman, the fact that he does not wear wooden shoes does not render my opinion invalid. To this Hume, in his turn, can reply: How do you know that 'Dutchmen wear wooden shoes'? – you have to observe separately (i) Dutchmen, (ii) wooden shoes. No, says Aristotle, why would this be so? Everybody knows that Dutchmen wear wooden shoes: it is a synthetic proposition a priori. To this Hume can only – as far as I see – retort: I do not see how such a proposition that is independent of phenomena can be used as a norm to qualify phenomena. Moreover, it is senseless to make inquiries (to come to Holland) if one has knowledge a priori (knows already all about Holland). Aristotle, however, can also attack Hume; he can say: Your point of view is that testing is necessary because we know by perception – that is, *independently* of one another – (i) whether somebody is a Dutchman; (ii) whether somebody is wearing wooden shoes. But does one know that qualifications-by-perception (i.e. by one and the same person) are not interdependent? That is nothing but an assumption: you not only believe a priori in the trustworthiness of perception – so do I – but you also believe a priori that there are such things as atomic impressions. The practice of daily life, as well as the practice of scientific inquiry, demonstrates that some universal propositions are valid: people use them as norms if they use their knowledge as a guide to action. To this last part of the argument Hume can answer: This only proves that beliefs are necessary in practice. But Aristotle will reply: I do not see what other criterion of validity can be given, if it does not refer to a course of action. Certainly, a certain course of action need not give the desired results; but that only shows that perception is not always reliable: people can – and often do – make mistakes.

Here I will close the discussion. I prefer to say that one cannot in general, *in abstracto*, once and for all, make the decision to agree either with 'Hume' or with 'Aristotle'. One has to decide afresh every time – in given circumstances and with an eye on facts on the one hand and the proposition in question on the other hand – to decide whether the proposition should be tested or whether it should be used as a norm to judge phenomena. To make this decision once and for all – that is, *in abstracto* – is irrational, for one thereby gives up the only foothold that is available.

There is one point of the foregoing discussion which I should like to stress: Hume's point of view rests on the assumption that '*there are*' primitive and elementary sensations; the point of view of Aristotle on the assumption that '*there are*' synthetic propositions a priori. This 'there are' is nonsense: man has to make incidental – but not irrational – decisions. On the one hand he has to use universal propositions in his inquiry into nature; if he uses them in this way they function as (relative) synthetic propositions a priori and are used as norms. On the other hand the inquirer has to test his propositions by observations; while doing so he puts his trust in simple sacrosanct descriptions. But again: neither laws nor descriptions 'are' by themselves unimpeachable. It is an illegitimate oversimplification to think so.

III. CONCLUSION

In the foregoing I have tried without any pretension to exhaustiveness to give some arguments against the traditional views about observation and description. One should not forget, I should like to add, that this view is itself an interpretation in terms of a model or picture of what is happening when we form a concept by simple apprehension and when we give a description of a situation. And one should remind oneself that though this picture of the process of observation and description may appeal to the imagination, it is only a picture, and that however 'clear' this picture may be, it can still be misleading.

In the following I shall not give such a picture but only try to analyse what we do; that is to say I shall try to analyse the interplay of concepts and universal propositions (or, more general: 'laws'). I will – on the basis of the practice of scientific inquiry – assume: (i) that we often presume *descriptions* to be valid, e.g. when we want to test laws; (ii) that we often presume *rules* to be valid, e.g. when we want to give a description by means of a description and a law.

(1) I stated above that neither laws nor descriptions are unimpeachable; I will now give a reason for this statement.

Wittgenstein advises us not to ask for the meaning (of a word) but to ask for the use, and states that 'the meaning is the use'. So far so good: a child has to learn the meaning of a word; that is to say: it has to learn how to use it. But – one can ask – at what point does the child become an expert himself? The answer can be (Cf. Van Ginneken, *Mysterie der menselijke taal*): At the moment at which his use of a word is in conformity with the use of this word by grown-ups. But are grown-ups experts? Are they not able to learn – do not words change their meaning? Certainly they do and certainly we can learn – scientific inquiry itself is a determinate endeavour to learn; it is itself a constant renewal of acquired knowledge (not only of the laws and theories but also of concepts).

One cannot simply wait to become an expert; one has to use words to be able to learn how to use them. That means that, at a certain moment, one says (without saying so): now I know how to use a word; I know whether a given description by means of a given word is true or false. But how can we know whether this use is right? (The correspondence theory of truth is here of no use.)

Concerning laws we can – mutatis mutandis – say the same things: we form and test them – but never conclusively. However: there comes a moment when we have to use them: how can we know whether they are valid?

One thing is clear: the position of concepts and laws is, in respect of their validity, the same: the one is not either more, or less, unimpeachable than the other. And this fact provides us with a standard; but not with an absolute one. Or more exactly: this fact enables us to learn how to use words and how to make laws.

(2) Before going farther into this question I want to give another argument – not based upon a picture – in favour of Hume's point of view.

A law of the form 'if A then B' enables us to give, by means of a description 'A', another description 'B'. If observation reveals to us that 'B' is false we have to make a decision: we can either say that the law is invalid or we can say that the description 'A' – based on observation – is false or again that 'B' is not false. The positions are not however symmetrical: to be able to state (not unfoundedly) that the law is invalid we must rely on observation – but only on observation. To be able to state (not unfoundedly) that a description is invalid we need firstly a law and secondly a description. Therefore observation is more fundamental than the institution of a law.

This argument seems conclusive, nevertheless I do not accept it: in practice we often reject an observation on the basis of a law and other instances (later on I will give instances of this). Moreover: we can turn the argument the other way round. We can say that the law enables us to compare – and thus test – the use of one word with the use of another: in virtue of the law either 'A' is false or 'B' is not false. That is: by making use of laws we can learn how to use words in such a way that they enable us (1) to give descriptions of phenomena, (2) to introduce words in such a way that laws can be valid. So my suggestion is that we need laws for the formation of concepts, and that it is nonsense and the root of much nonsense to suppose that one could rely simply and solely on observation (in the sense of simple apprehension). Concepts have to be formed by means of laws. On the other hand there is always the possibility of relying on our expertism in the use of words to form and test laws on this basis. (Which of them one should do this to cannot be decided *in abstracto* and in general: we are neither recording machines which are fed bits of information nor programmed computers.)

To return to my statement about how it is possible to learn to use words and to form laws, I will give a simile and compare a *mother* who teaches her child how to use a word, with a *law* that is used as a norm to teach us how to use words (in the plural). In this case the mother as well as the law have authority – but the law leaves us freedom of choice: we have possibly to reconsider our use of 'A' or of 'B' or of both (one word is the sharpening stone for the other). A second remark: if we obey the law it becomes automatically valid and can be stated (instituted) explicitly. Another remark: the problem of induction is not how to institute valid laws (presupposing words with fixed meanings); it is how to form and institute both words which can be used as descriptions *and* laws that are valid. But so stated the problem of induction is simply the problem of scientific inquiry as a whole.

Let me again give a simile: a law enables us – given one description – to give another description; in this sense the law is a road-between. And who can say what will be suitable places for the building of villages without taking into consideration the possibility of constructing roads that connect them, or, vice-versa, what will be good roads without taking into consideration the question whether these roads suitably connect certain villages? Moreover, are not cross-roads themselves often suitable places for the building of villages, or, vice versa, are not cross-roads often found in villages because of the necessity to connect one village with many other

villages? The only thing we can say is that the 'suitability' of a place is never determined solely in respect to the natural qualities of this place itself; the possibility of connecting this place with others has to be taken into account.
(3) In the preceding remarks only words designating qualities and things were mentioned. Now the traditional theory loses much of its 'clearness' if one tries to take numerical descriptions into account. It is easy enough to state that conceptions are formed by simple apprehension and abstraction and that passive observation – and only passive observation – supports the experiential proposition (and – by doing so – gives it its sacrosanct status). But it is not easy (i) to see what is the meaning of a numerical description; (ii) to imagine how we form by abstraction numbers as for instance two-hundred-and-seventy-three.
Let me state (1) that I think that the notion 'number' is based not upon abstraction but upon construction (the 'and so on' is an essential feature); (2) that numbers such as 'five', 'six', 'seven' ... have in virtue of this construction an intra-systematic meaning; this implies that the experiential proposition 'this packet contains 17 cigarettes' automatically entails: 'this packet contains more than 10 cigarettes' – in short: when we give a description by means of numbers we know that other descriptions are entailed; (3) that we can only find out that a packet contains 17 cigarettes by counting them, that is to say: these descriptions are based upon operations and not on 'passive observation'; (4) that by performing this operation we adhere to certain formal rules (see (2): intra-systematic meaning).
As far as I can see no theory of knowledge – knowledge that includes our physical theories – can in our day refrain from declaring itself on this point. One has to make a decision: one can adhere to the traditional opinion and try to account for numbers either in terms of descriptions (but how?) or by saying that such descriptions have no objective content because they are not based upon abstraction. Now one should do neither: it seems to me that exactly these descriptions by means of numbers can make clear to us what we do when we qualify situations by giving 'normal' descriptions. Or to state it otherwise: there is in principle no disparity between the two types of descriptions. When we give a description such as, e.g., 'this is red' we in no sense passively copy a particular situation but we perform an operation and in doing so adhere to certain rules.
Let me give an illustration. When a mother teaches her child the meaning of the word 'warm' – teaches her child how to use it – she sometimes says

'this is warm', or – if the child tries to use the word for a cold thing – says: 'no – this is not-warm'. And so the child, in the long run, becomes an expert: when asked, it can say of things (within a certain range, in certain circumstances) whether they are 'warm' or 'not-warm'. When these distinctions are made it may be necessary to make sharper distinctions, e.g. in the realm of the things that are 'not-warm' to speak of 'lukewarm' and 'cold'. And so, by being taught how to speak, (i) the child learns to make distinctions; (ii) he becomes an expert in the use of different words (not: first a warm thing is christened, then – and independently – a lukewarm or a cold thing; the use of the words 'warm', 'lukewarm', 'cold' is learned together). This means that in order to give a description, the child has to make a choice; in making this choice it has to adhere to rules such as 'a thing is warm, or lukewarm, or cold'. These rules are valid formal rules; by looking at actually given descriptions one by one nothing can be seen of these laws: it is only by looking at all these descriptions together that one can conclude that they play a role in the actual decisions (in this sense they are valid). Or to put it otherwise: these laws are in the case of relatively simple qualitative statements created at the same time as the meaning of the words; or: the extra-systematic and the intra-systematic meaning of these words is learned at the same time. And therefore we can say that whoever says that 'a warm thing is cold', 'does not know what he is saying' or 'speaks nonsense' – we do not say that somebody who does do so is uttering a false statement. To contrast this point of view with the traditional theory: on the basis of simple apprehension one has to explain why we are sure that 'warm things are not cold' but not that 'warm things are not blue'. And how could that be done when warm, cold and blue are independent sensations or qualities?

(4) There is another remark to be made concerning the introduction of concepts in physical inquiry. Here conceptions are introduced on the basis of other concepts assuming the validity of laws – certainly not by simple apprehension and abstraction. I will give an illustration.

(4.1) Ohm's law, $V = i.r$, is an empirical law; it can be found for instance by measuring corresponding values of V and i. These pairs of values can be plotted in a graph and it appears that a straight line can be drawn through these points. In this way it is clear that $V = i.r$.

(4.2) We are accustomed to saying that an empirical law may in principle be shown to be false; this means that such an empirical or inductive law

is only valid 'bis auf weiteres' (Kant). Or again: we cannot neglect the possibility that some day phenomena will be observed which are not covered by the law. By doing so we neglect an essential feature: in the normal routine of physical inquiry such non-fitting facts are nearly always present (you have only to look at graphs in papers of experimental physicists to know that this statement is true). Therefore it is misleading to direct the attention to the possibility of falsification or to possible future non-fitting facts. Such facts are nearly always already available. Nevertheless the validity of the laws is assumed: this means that we cannot say: a law is valid if, and only if, it covers all observed phenomena. This in turn implies that 'falsity' is not a clear-cut notion.
Whenever an experimental physicist finds such non-fitting facts he is in the habit of saying: these facts are not facts but faults. And if one asks him how he knows this, he will answer: 'Because they do not fit into the general picture. (Compare this attitude with the attitude of a judge who, solely on the basis of the evidence, can accuse some witness of perjury.)
In short: one cannot say that a law or a theory is valid if, and only if, it covers all experiential data; or again: induction (as the operation by which we institute laws) is neither simply and only a question of experiment nor a question of deduction. One always has to make a decision to institute a law; this decision however is made in view of the material on hand – that material and only that provides the standard or norm by which a judgement is made. Therefore this decision itself (the formulation of the law on basis of the material on hand) is neither a logical conclusion (some premisses are abolished in the act of judgement) nor an arbitrary act (it is a decision in view of the material on hand). (In virtue of the fact that in actual induction 'facts' are rendered invalid, I think that the search for a 'logic of induction' is a rather academic affair: one has – in the act of judgement – to determine what is acceptable as 'evidence'.)
(4.3) To return to (i) the institution of new conceptions; (ii) the illustration given above ($V = i.r$): if we had not been able to draw a straight line through the points (corresponding pairs of values of V and i) we would not have been able to introduce the concept 'resistance'; this concept derives its meaning and its validity from Ohm's law (and because this law can only be instituted by abolishing certain 'facts' this concept is not a simple 'construct'). In other words: the institution of the law entails the institution of a concept which can be used to give descriptions of

phenomena, e.g. 'The resistance of this wire is 3 Ω'. Two remarks should be made: (i) to give this description we have to perform operations (we have given an operational description of the concept 'resistance'; but again: this definition makes sense if, and only if, Ohm's law is valid); (ii) the value of the resistance (3 Ω) is, of course, independent of the value for the potential difference we choose to apply to the wire; it is a property 'of the wire itself' – but only in virtue of Ohm's law (it is not only 'a' quotient of 'a' value of V and 'a' value of i). To repeat: new concepts can be generated by a combination of observing and thinking (we need our mind to see). Scientific knowledge has enormous self-propelling power because this sort of knowledge (in the sense of knowledge of laws) can be used as a means to create – in an objective sense – concepts which can be used to give descriptions of phenomena; acquired knowledge enables us not only to direct our activities in everyday life but also – *and this is done in inquiry* – to see deeper and farther than was possible beforehand. In this double sense the physicist H. A. Kramers has said: 'How do I know that I have understood? This I know in virtue of the fact that I can now act in such a way as was not possible beforehand'.

(4.4) Once the concept 'resistance' has been instituted by an operational definition, one can, for instance, measure resistance as a function of temperature. In doing so one again finds a law. Now a sceptic can try to falsify this law by looking for pairs of values for resistance and temperature which do not fit into this law – but to do so, he has to presuppose that Ohm's law is valid. I mention this point because it makes clear that the possibility of falsifying a law depends on the assumed validity of descriptions of a certain type which in their turn depend on a law: the 'objective value' of concepts is complementary to the validity of laws.

Now this implies that one can state that somewhere apodictive decisions have to be made. In doing so one can assume either that some concepts are sacrosanct or that some laws have an a priori status; this means one can assume that descriptions of a certain type are valid or that certain laws are valid. The first choice is made by empiricists of Humean origin – the latter by rationalists of the Aristotelian or Cartesian tradition who are looking for 'first principles'. But the scientist does not make this kind of decision: for him neither concepts nor laws are sacrosanct: he tries to form concepts, to learn how to make observations and to use words by means of laws – he tries to formulate laws by making observations and using words; moreover inquiry itself is a process of transformation and readjustment. What is true in the empiricist's point of view is that one can only acquire knowledge by making observations – what is true in the point of view of the rationalist is that one can only acquire knowledge by stating laws; the scientist tries to see deeper and farther by using his brains and making rational decisions – he tries to learn and understand more and more by using his eyes.

(5) I return to the initial question: is knowledge a model, picture or image of nature? And is it true in virtue of its conformity to nature? This question is certainly very vague – one can ask what can be meant by it. However, I prefer not to try to make this meaning any clearer – what interests me is the suggestion implied in the question. And this suggestion is that true knowledge has to be fixed, unalterable; true knowledge has – as an image of nature – to be fixed as long as 'nature is what it is'. Renewal and re-adjustment itself would then be symptoms of the inadequacy of the acquired knowledge. Or to put it otherwise: if knowledge is an image of nature we should – I return to the third Commandment – 'bow down to it and serve it'; i.e., it could not be used but should be served.

Now there is another (not transcendent) criterion for the validity of knowledge. When Kant called logic, mathematics and physics 'examples of science' he did so in virtue of their steadfast progress ('sicheren Gang') ('kein blosses Herumtapfen'); this criterion can and should be applied.

I think that no student of physics will deny that acquired knowledge initiates and directs inquiry; inquiry which in its turn creates new knowledge and transforms old knowledge. This means that knowledge itself is not an inquiry-frustrating image, but, on the contrary, an *instrument* that enables us to see what we cannot see without its help. It directs our eyes towards phenomena which, in their turn, make it possible to acquire new knowledge or make it necessary to transform our initial knowledge. In this sense knowledge itself creates the standards by which its value should be measured.

In short: because knowledge enables us to see what we could not see, because it has in this sense 'realizing power' and because it enables us to make steadfast progress and is self-correcting – because it can do all these things, therefore, it should be called 'true'. But for the same reasons it should not be called a 'model'.

University of Leyden